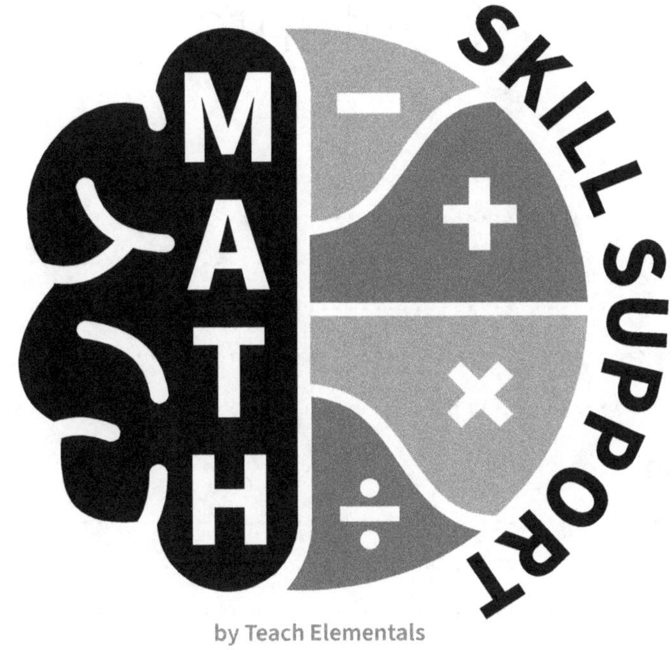

Adding and Subtracting to 20

ISBN: 978-1-964695-02-0

Published by Teach Elementals, LLC.

Contents

Section A: Addition

Adding 0 (2 sets) 1
Adding 1 (2 sets) 3
Adding 10 (2 sets) 5
Adding up to 10 (5 sets) 7
Adding up to 20 (21 sets) 12

Section B: Subtraction

Subtracting 0 (2 sets) 33
Subtracting 1 (2 sets) 35
Subtracting 10 (2 sets) 37
Subtracting within 10 (5 sets) 39
Subtracting within 20 (21 sets) 44

Section C: Addition and Subtraction

Mixed Review: Evens Only (5 sets) 65
Mixed Review: Odds Only (5 sets) 70
Mixed Review: Adding and Subtracting 0-20 (26 sets) 75

Answer Keys 101

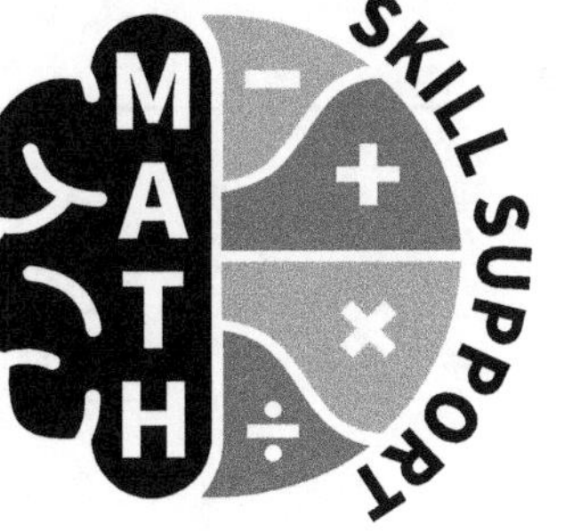

Name: ______________________ Date: ______________

DAY 1

ADDING 0, SET A

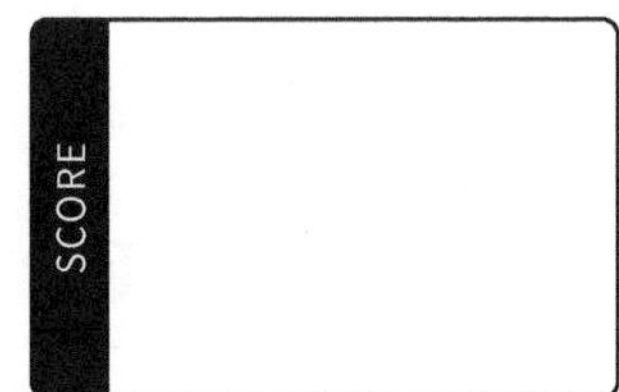

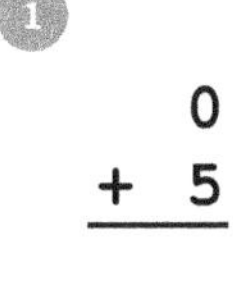

				1) 0 + 5	2) 0 +10	3) 0 +19	4) 0 +11
5) 7 + 0	6) 0 + 1	7) 0 +15	8) 0 + 6	9) 4 + 0	10) 13 + 0	11) 0 + 2	12) 2 + 0
13) 0 +20	14) 0 + 7	15) 0 +13	16) 20 + 0	17) 3 + 0	18) 0 + 3	19) 12 + 0	20) 0 +18
21) 9 + 0	22) 0 +12	23) 8 + 0	24) 14 + 0	25) 0 +16	26) 10 + 0	27) 16 + 0	28) 0 + 4
29) 17 + 0	30) 19 + 0	31) 15 + 0	32) 0 + 9	33) 6 + 0	34) 0 + 8	35) 11 + 0	36) 5 + 0
37) 18 + 0	38) 0 +17	39) 0 + 0	40) 1 + 0	41) 0 +14	42) 10 + 0	43) 7 + 0	44) 13 + 0
45) 0 + 7	46) 3 + 0	47) 0 +13	48) 0 + 8	49) 0 +20	50) 17 + 0	51) 0 + 3	52) 16 + 0
53) 0 + 6	54) 20 + 0	55) 0 + 1	56) 0 + 5	57) 0 + 0	58) 0 +11	59) 0 + 9	60) 0 +18

DAY 2

ADDING 0, SET B

Name: ____________________ Date: ____________

				1. 0 + 16	2. 0 + 9	3. 0 + 10	4. 14 + 0
5. 9 + 0	6. 0 + 4	7. 18 + 0	8. 20 + 0	9. 0 + 8	10. 4 + 0	11. 0 + 0	12. 0 + 14
13. 10 + 0	14. 0 + 3	15. 0 + 6	16. 6 + 0	17. 17 + 0	18. 0 + 20	19. 0 + 12	20. 0 + 5
21. 13 + 0	22. 0 + 13	23. 0 + 1	24. 15 + 0	25. 5 + 0	26. 3 + 0	27. 16 + 0	28. 0 + 18
29. 8 + 0	30. 0 + 2	31. 19 + 0	32. 1 + 0	33. 2 + 0	34. 0 + 17	35. 0 + 15	36. 0 + 11
37. 12 + 0	38. 7 + 0	39. 0 + 7	40. 11 + 0	41. 0 + 19	42. 0 + 12	43. 12 + 0	44. 0 + 4
45. 6 + 0	46. 7 + 0	47. 3 + 0	48. 11 + 0	49. 0 + 11	50. 0 + 8	51. 4 + 0	52. 15 + 0
53. 0 + 14	54. 14 + 0	55. 0 + 2	56. 0 + 18	57. 10 + 0	58. 0 + 16	59. 20 + 0	60. 17 + 0

ADDING 1, SET A

Name: ______________________ Date: ______________

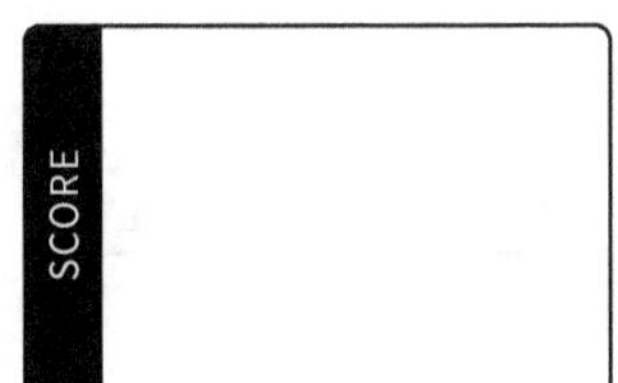

1. $17 + 1$
2. $9 + 1$
3. $1 + 3$
4. $1 + 7$
5. $0 + 1$
6. $1 + 9$
7. $16 + 1$
8. $4 + 1$
9. $1 + 1$
10. $18 + 1$
11. $8 + 1$
12. $10 + 1$
13. $1 + 11$
14. $1 + 12$
15. $1 + 13$
16. $13 + 1$
17. $1 + 16$
18. $6 + 1$
19. $2 + 1$
20. $1 + 0$
21. $1 + 6$
22. $1 + 10$
23. $7 + 1$
24. $1 + 18$
25. $3 + 1$
26. $11 + 1$
27. $1 + 5$
28. $1 + 4$
29. $15 + 1$
30. $14 + 1$
31. $1 + 19$
32. $1 + 14$
33. $12 + 1$
34. $1 + 17$
35. $1 + 2$
36. $19 + 1$
37. $1 + 8$
38. $5 + 1$
39. $1 + 15$
40. $1 + 12$
41. $1 + 15$
42. $1 + 6$
43. $3 + 1$
44. $9 + 1$
45. $17 + 1$
46. $15 + 1$
47. $0 + 1$
48. $1 + 4$
49. $1 + 3$
50. $1 + 16$
51. $7 + 1$
52. $4 + 1$
53. $10 + 1$
54. $1 + 2$
55. $1 + 18$
56. $1 + 10$
57. $1 + 1$
58. $1 + 8$
59. $1 + 0$
60. $12 + 1$

Name: ____________________ Date: ____________

				1 17 + 1	**2** 1 +19	**3** 1 +10	**4** 1 + 4
5 1 +14	**6** 11 + 1	**7** 1 +11	**8** 1 +13	**9** 1 + 0	**10** 6 + 1	**11** 18 + 1	**12** 15 + 1
13 1 +15	**14** 19 + 1	**15** 2 + 1	**16** 1 + 7	**17** 1 +16	**18** 14 + 1	**19** 3 + 1	**20** 1 + 2
21 5 + 1	**22** 4 + 1	**23** 12 + 1	**24** 1 + 1	**25** 1 + 9	**26** 1 + 5	**27** 1 + 6	**28** 10 + 1
29 1 + 8	**30** 7 + 1	**31** 0 + 1	**32** 8 + 1	**33** 9 + 1	**34** 1 +17	**35** 1 +12	**36** 16 + 1
37 13 + 1	**38** 1 +18	**39** 1 + 3	**40** 2 + 1	**41** 1 +12	**42** 15 + 1	**43** 1 +13	**44** 1 +19
45 1 +10	**46** 1 + 3	**47** 0 + 1	**48** 1 +15	**49** 4 + 1	**50** 9 + 1	**51** 6 + 1	**52** 18 + 1
53 1 + 9	**54** 1 + 8	**55** 8 + 1	**56** 10 + 1	**57** 7 + 1	**58** 1 + 5	**59** 1 +17	**60** 5 + 1

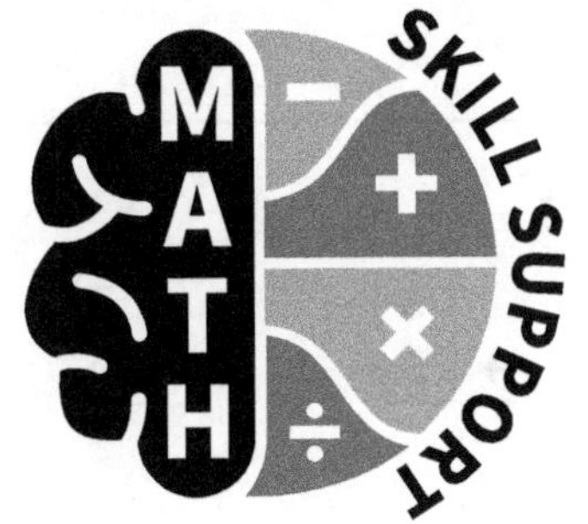

Name: ____________________ Date: ____________________

SCORE

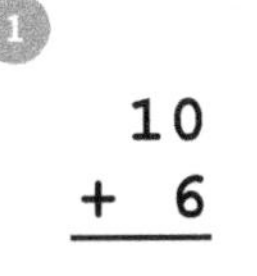

				1. 10 + 6	2. 1 + 10	3. 9 + 10	4. 10 + 1
5. 2 + 10	6. 10 + 10	7. 10 + 8	8. 6 + 10	9. 0 + 10	10. 10 + 4	11. 5 + 10	12. 7 + 10
13. 8 + 10	14. 10 + 5	15. 10 + 10	16. 10 + 2	17. 10 + 9	18. 10 + 3	19. 3 + 10	20. 4 + 10
21. 10 + 7	22. 10 + 0	23. 10 + 10	24. 10 + 9	25. 10 + 1	26. 10 + 6	27. 10 + 4	28. 10 + 5
29. 2 + 10	30. 9 + 10	31. 10 + 0	32. 7 + 10	33. 5 + 10	34. 1 + 10	35. 6 + 10	36. 10 + 3
37. 10 + 8	38. 10 + 10	39. 8 + 10	40. 0 + 10	41. 3 + 10	42. 10 + 2	43. 4 + 10	44. 10 + 7
45. 5 + 10	46. 9 + 10	47. 1 + 10	48. 10 + 3	49. 10 + 10	50. 10 + 4	51. 10 + 2	52. 10 + 8
53. 10 + 9	54. 3 + 10	55. 4 + 10	56. 10 + 5	57. 10 + 1	58. 7 + 10	59. 10 + 7	60. 2 + 10

DAY 6

ADDING 10, SET B

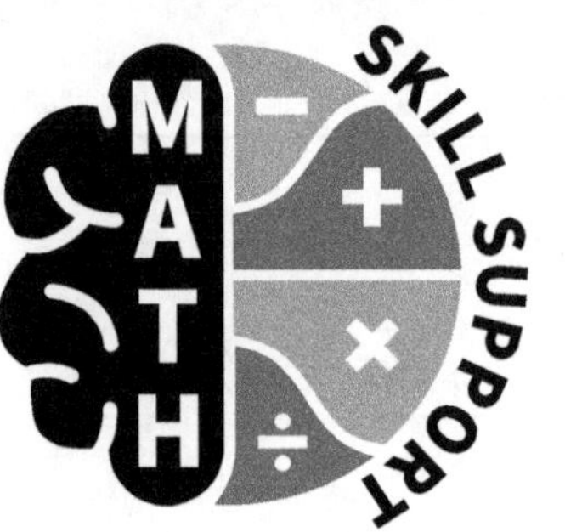

Name: ____________________ Date: ____________

1. 10 + 7	2. 10 + 6	3. 10 + 10	4. 3 + 10

5. 2 + 10	6. 1 + 10	7. 10 + 8	8. 7 + 10	9. 10 + 3	10. 0 + 10	11. 4 + 10	12. 10 + 0
13. 10 + 5	14. 10 + 9	15. 5 + 10	16. 10 + 10	17. 9 + 10	18. 6 + 10	19. 10 + 4	20. 8 + 10
21. 10 + 2	22. 10 + 1	23. 8 + 10	24. 1 + 10	25. 10 + 8	26. 10 + 6	27. 10 + 7	28. 10 + 3
29. 10 + 10	30. 6 + 10	31. 10 + 5	32. 10 + 4	33. 2 + 10	34. 10 + 10	35. 10 + 9	36. 10 + 1
37. 10 + 0	38. 9 + 10	39. 5 + 10	40. 3 + 10	41. 10 + 2	42. 4 + 10	43. 0 + 10	44. 7 + 10
45. 10 + 2	46. 10 + 8	47. 10 + 4	48. 3 + 10	49. 10 + 7	50. 7 + 10	51. 0 + 10	52. 10 + 10
53. 10 + 0	54. 10 + 3	55. 5 + 10	56. 2 + 10	57. 10 + 9	58. 9 + 10	59. 1 + 10	60. 6 + 10

MATH SKILL SUPPORT

Name: ______________________ Date: ______________________

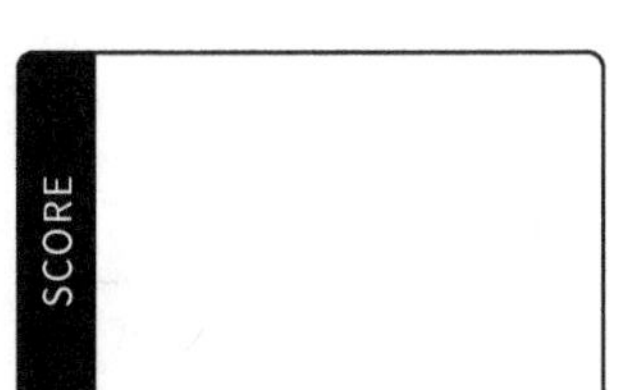

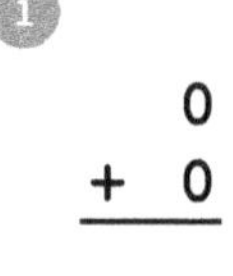

ADDING UP TO 10, SET A

				1. $0 + 0$	2. $0 + 1$	3. $0 + 2$	4. $0 + 3$
5. $0 + 4$	6. $0 + 5$	7. $0 + 6$	8. $0 + 7$	9. $0 + 8$	10. $0 + 9$	11. $0 + 10$	12. $1 + 0$
13. $1 + 1$	14. $1 + 2$	15. $1 + 3$	16. $1 + 4$	17. $1 + 5$	18. $1 + 6$	19. $1 + 7$	20. $1 + 8$
21. $1 + 9$	22. $2 + 0$	23. $2 + 1$	24. $2 + 2$	25. $2 + 3$	26. $2 + 4$	27. $2 + 5$	28. $2 + 6$
29. $2 + 7$	30. $2 + 8$	31. $3 + 0$	32. $3 + 1$	33. $3 + 2$	34. $3 + 3$	35. $3 + 4$	36. $3 + 5$
37. $3 + 6$	38. $3 + 7$	39. $4 + 0$	40. $4 + 1$	41. $4 + 2$	42. $4 + 3$	43. $4 + 4$	44. $4 + 5$
45. $4 + 6$	46. $5 + 0$	47. $5 + 1$	48. $5 + 2$	49. $5 + 3$	50. $5 + 4$	51. $5 + 5$	52. $6 + 0$
53. $6 + 1$	54. $6 + 2$	55. $6 + 3$	56. $6 + 4$	57. $7 + 0$	58. $7 + 1$	59. $7 + 2$	60. $7 + 3$

MATH SKILL SUPPORT

Name: ____________________ Date: ____________

SCORE

				1. 1 + 6	2. 0 + 8	3. 0 + 1	4. 0 + 3
5. 8 + 2	6. 3 + 7	7. 3 + 1	8. 1 + 2	9. 4 + 3	10. 4 + 2	11. 1 + 5	12. 3 + 6
13. 1 + 4	14. 4 + 6	15. 3 + 2	16. 0 + 4	17. 10 + 0	18. 8 + 1	19. 3 + 5	20. 2 + 7
21. 0 + 0	22. 5 + 3	23. 9 + 0	24. 4 + 5	25. 6 + 2	26. 6 + 0	27. 7 + 0	28. 7 + 1
29. 3 + 0	30. 2 + 1	31. 5 + 4	32. 2 + 3	33. 2 + 0	34. 0 + 10	35. 0 + 2	36. 5 + 5
37. 0 + 5	38. 1 + 1	39. 2 + 2	40. 1 + 3	41. 1 + 8	42. 4 + 0	43. 3 + 4	44. 4 + 4
45. 8 + 0	46. 5 + 2	47. 9 + 1	48. 2 + 5	49. 6 + 4	50. 7 + 2	51. 0 + 7	52. 6 + 1
53. 6 + 3	54. 7 + 3	55. 2 + 8	56. 2 + 6	57. 1 + 0	58. 0 + 6	59. 1 + 7	60. 3 + 3

Name: ______________________ Date: ______________

				1. $7 + 1$	2. $2 + 7$	3. $0 + 5$	4. $2 + 4$
5. $4 + 0$	6. $4 + 2$	7. $2 + 6$	8. $2 + 3$	9. $8 + 0$	10. $1 + 2$	11. $1 + 4$	12. $4 + 5$
13. $6 + 2$	14. $0 + 2$	15. $8 + 1$	16. $2 + 0$	17. $3 + 4$	18. $1 + 5$	19. $0 + 10$	20. $9 + 1$
21. $2 + 5$	22. $0 + 9$	23. $5 + 2$	24. $0 + 4$	25. $0 + 7$	26. $10 + 0$	27. $5 + 4$	28. $1 + 7$
29. $8 + 2$	30. $3 + 5$	31. $0 + 0$	32. $6 + 4$	33. $4 + 4$	34. $0 + 8$	35. $3 + 7$	36. $0 + 3$
37. $5 + 1$	38. $6 + 3$	39. $3 + 6$	40. $1 + 8$	41. $3 + 2$	42. $5 + 3$	43. $1 + 6$	44. $1 + 9$
45. $4 + 6$	46. $7 + 2$	47. $7 + 0$	48. $6 + 0$	49. $2 + 2$	50. $6 + 1$	51. $2 + 1$	52. $9 + 0$
53. $7 + 3$	54. $2 + 8$	55. $4 + 3$	56. $3 + 3$	57. $4 + 1$	58. $1 + 3$	59. $3 + 0$	60. $0 + 6$

MATH SKILL SUPPORT

Name: ____________________ Date: ____________

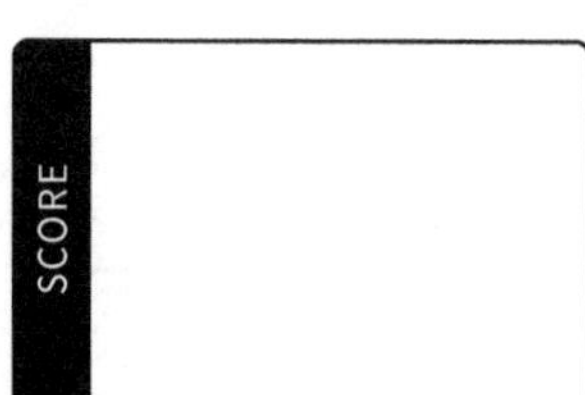

				1. 3 + 1	2. 2 + 8	3. 1 + 2	4. 8 + 1
5. 0 + 1	6. 2 + 4	7. 1 + 1	8. 2 + 7	9. 6 + 3	10. 0 + 3	11. 4 + 5	12. 2 + 3
13. 3 + 5	14. 6 + 0	15. 4 + 2	16. 7 + 0	17. 7 + 1	18. 0 + 5	19. 1 + 0	20. 0 + 9
21. 1 + 7	22. 5 + 5	23. 1 + 5	24. 0 + 6	25. 8 + 0	26. 1 + 4	27. 4 + 4	28. 6 + 1
29. 0 + 4	30. 2 + 5	31. 5 + 2	32. 0 + 0	33. 2 + 2	34. 2 + 0	35. 2 + 6	36. 0 + 7
37. 6 + 2	38. 6 + 4	39. 3 + 0	40. 0 + 10	41. 3 + 6	42. 4 + 6	43. 9 + 0	44. 3 + 7
45. 10 + 0	46. 3 + 4	47. 5 + 1	48. 5 + 4	49. 1 + 3	50. 4 + 1	51. 3 + 3	52. 1 + 6
53. 7 + 2	54. 2 + 1	55. 0 + 2	56. 9 + 1	57. 0 + 8	58. 5 + 3	59. 4 + 3	60. 3 + 2

MATH SKILL SUPPORT

Name: ______________________ Date: ______________

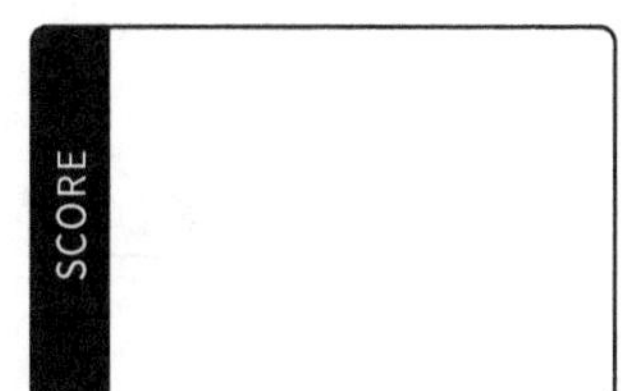

				1. 6 + 4	2. 3 + 2	3. 5 + 0	4. 9 + 1
5. 0 + 6	6. 10 + 0	7. 7 + 1	8. 6 + 2	9. 2 + 3	10. 1 + 9	11. 1 + 3	12. 1 + 5
13. 8 + 0	14. 3 + 1	15. 3 + 0	16. 6 + 3	17. 2 + 1	18. 5 + 4	19. 0 + 3	20. 0 + 9
21. 5 + 1	22. 0 + 1	23. 0 + 5	24. 1 + 4	25. 9 + 0	26. 2 + 6	27. 1 + 8	28. 8 + 2
29. 0 + 4	30. 7 + 0	31. 3 + 4	32. 1 + 7	33. 4 + 4	34. 2 + 4	35. 0 + 2	36. 2 + 5
37. 0 + 7	38. 4 + 2	39. 3 + 3	40. 2 + 2	41. 1 + 0	42. 0 + 0	43. 4 + 3	44. 7 + 2
45. 2 + 7	46. 4 + 0	47. 2 + 0	48. 3 + 6	49. 1 + 1	50. 5 + 3	51. 5 + 2	52. 4 + 6
53. 3 + 7	54. 1 + 6	55. 8 + 1	56. 6 + 1	57. 1 + 2	58. 0 + 10	59. 6 + 0	60. 0 + 8

MATH SKILL SUPPORT

Name: ____________________ Date: ____________

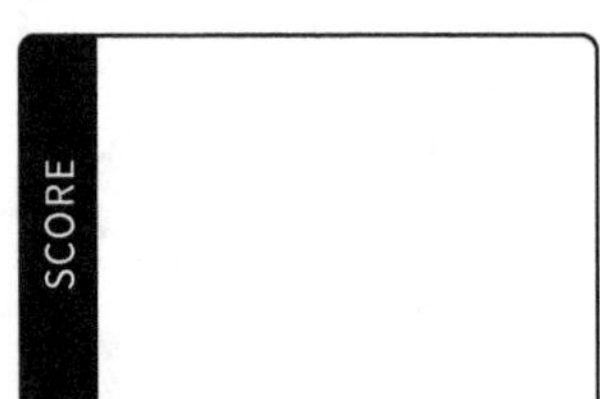
SCORE

1	2	3	4
0 + 15	1 + 14	9 + 9	3 + 14

5	6	7	8	9	10	11	12
4 + 6	6 + 12	15 + 4	19 + 1	4 + 12	3 + 8	12 + 7	7 + 0

13	14	15	16	17	18	19	20
7 + 5	11 + 1	11 + 6	18 + 0	3 + 7	3 + 0	3 + 13	1 + 12

21	22	23	24	25	26	27	28
7 + 1	5 + 1	9 + 4	17 + 3	4 + 5	2 + 8	6 + 11	9 + 5

29	30	31	32	33	34	35	36
5 + 10	15 + 3	13 + 3	16 + 0	18 + 1	0 + 17	2 + 6	2 + 1

37	38	39	40	41	42	43	44
1 + 10	5 + 9	2 + 4	12 + 8	4 + 16	7 + 6	13 + 1	7 + 10

45	46	47	48	49	50	51	52
11 + 8	11 + 0	9 + 3	8 + 5	14 + 0	1 + 19	0 + 19	11 + 9

53	54	55	56	57	58	59	60
11 + 7	6 + 7	20 + 0	2 + 17	17 + 1	1 + 11	14 + 2	6 + 5

MATH SKILL SUPPORT

Name: ______________________ Date: ______________

DAY 13

ADDING UP TO 20, SET B

				1. $9 + 2$	2. $8 + 8$	3. $17 + 2$	4. $2 + 16$
5. $5 + 11$	6. $13 + 5$	7. $1 + 2$	8. $0 + 20$	9. $10 + 7$	10. $0 + 7$	11. $1 + 13$	12. $15 + 2$
13. $1 + 18$	14. $11 + 4$	15. $5 + 3$	16. $2 + 15$	17. $6 + 8$	18. $9 + 11$	19. $5 + 13$	20. $6 + 4$
21. $1 + 7$	22. $3 + 12$	23. $1 + 17$	24. $2 + 9$	25. $4 + 1$	26. $0 + 4$	27. $8 + 6$	28. $1 + 9$
29. $5 + 6$	30. $6 + 9$	31. $3 + 1$	32. $0 + 13$	33. $4 + 4$	34. $3 + 2$	35. $5 + 8$	36. $0 + 2$
37. $2 + 11$	38. $3 + 16$	39. $13 + 6$	40. $13 + 2$	41. $7 + 2$	42. $1 + 8$	43. $3 + 4$	44. $6 + 1$
45. $8 + 1$	46. $7 + 12$	47. $8 + 11$	48. $4 + 2$	49. $6 + 13$	50. $12 + 2$	51. $2 + 3$	52. $4 + 0$
53. $0 + 8$	54. $7 + 11$	55. $4 + 7$	56. $3 + 6$	57. $12 + 3$	58. $13 + 4$	59. $8 + 2$	60. $12 + 6$

MATH SKILL SUPPORT

Name: ____________________ Date: ____________________

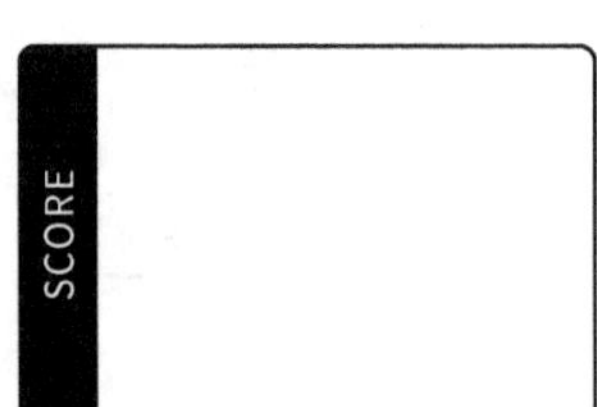
SCORE

				1. 2 +18	2. 1 +15	3. 10 + 8	4. 12 + 4
5. 12 + 1	6. 9 + 8	7. 2 + 7	8. 11 + 3	9. 6 + 2	10. 11 + 2	11. 0 +11	12. 7 + 3
13. 2 +10	14. 3 + 3	15. 7 + 7	16. 10 + 2	17. 0 + 6	18. 0 + 9	19. 0 + 3	20. 0 + 5
21. 1 + 6	22. 16 + 4	23. 4 +11	24. 0 +10	25. 0 +14	26. 1 + 1	27. 19 + 0	28. 8 + 3
29. 0 +16	30. 6 + 3	31. 10 + 3	32. 5 + 4	33. 8 +10	34. 14 + 1	35. 12 + 5	36. 10 +10
37. 3 +15	38. 9 + 0	39. 13 + 7	40. 14 + 6	41. 6 +14	42. 13 + 0	43. 5 + 0	44. 5 +14
45. 3 +11	46. 4 +15	47. 1 + 3	48. 0 +12	49. 14 + 5	50. 6 + 0	51. 2 + 0	52. 0 + 1
53. 3 +17	54. 5 +15	55. 2 + 2	56. 18 + 2	57. 0 +18	58. 4 +13	59. 8 + 9	60. 16 + 3

MATH SKILL SUPPORT

Name: ______________________ Date: ______________

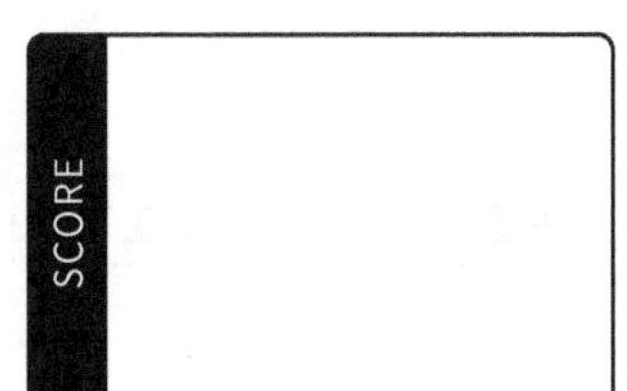
SCORE

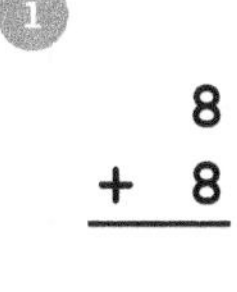
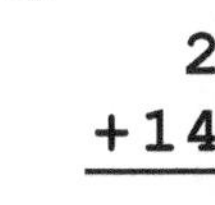
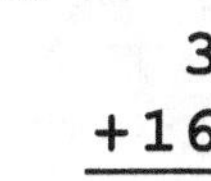

				1. 8 + 8	2. 2 + 14	3. 1 + 14	4. 3 + 16
5. 6 + 4	6. 2 + 7	7. 11 + 2	8. 0 + 4	9. 1 + 1	10. 6 + 13	11. 9 + 8	12. 5 + 3
13. 1 + 15	14. 18 + 2	15. 14 + 1	16. 1 + 19	17. 12 + 4	18. 1 + 18	19. 4 + 13	20. 0 + 11
21. 1 + 4	22. 5 + 9	23. 7 + 2	24. 2 + 18	25. 2 + 1	26. 0 + 2	27. 19 + 1	28. 17 + 0
29. 14 + 6	30. 17 + 3	31. 1 + 0	32. 2 + 11	33. 4 + 0	34. 8 + 4	35. 10 + 2	36. 6 + 6
37. 4 + 8	38. 5 + 7	39. 2 + 16	40. 18 + 1	41. 7 + 8	42. 13 + 6	43. 0 + 7	44. 14 + 2
45. 2 + 5	46. 3 + 1	47. 4 + 15	48. 11 + 1	49. 11 + 8	50. 6 + 10	51. 12 + 1	52. 1 + 11
53. 3 + 15	54. 7 + 3	55. 8 + 2	56. 11 + 6	57. 0 + 10	58. 19 + 0	59. 17 + 2	60. 1 + 5

SKILL SUPPORT MATH

Name: ____________________ Date: ____________

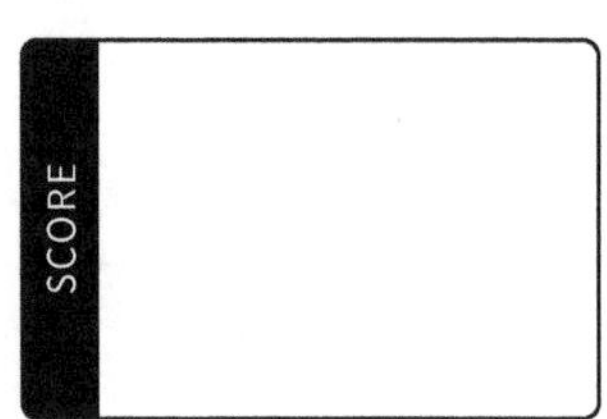
SCORE

				(1) 7 + 10	(2) 3 + 17	(3) 7 + 13	(4) 0 + 12
(5) 14 + 4	(6) 3 + 2	(7) 6 + 0	(8) 12 + 8	(9) 5 + 13	(10) 12 + 2	(11) 3 + 7	(12) 6 + 8
(13) 5 + 2	(14) 2 + 15	(15) 3 + 6	(16) 10 + 1	(17) 1 + 7	(18) 20 + 0	(19) 12 + 7	(20) 11 + 9
(21) 6 + 2	(22) 5 + 12	(23) 8 + 0	(24) 8 + 6	(25) 3 + 9	(26) 16 + 1	(27) 3 + 3	(28) 15 + 1
(29) 0 + 15	(30) 9 + 0	(31) 4 + 11	(32) 13 + 4	(33) 11 + 7	(34) 1 + 13	(35) 4 + 14	(36) 5 + 8
(37) 0 + 20	(38) 9 + 3	(39) 3 + 0	(40) 2 + 2	(41) 1 + 12	(42) 13 + 3	(43) 16 + 2	(44) 7 + 9
(45) 8 + 3	(46) 5 + 4	(47) 13 + 0	(48) 8 + 11	(49) 4 + 16	(50) 0 + 0	(51) 2 + 12	(52) 17 + 1
(53) 8 + 1	(54) 3 + 4	(55) 16 + 3	(56) 13 + 1	(57) 8 + 12	(58) 2 + 13	(59) 16 + 0	(60) 0 + 18

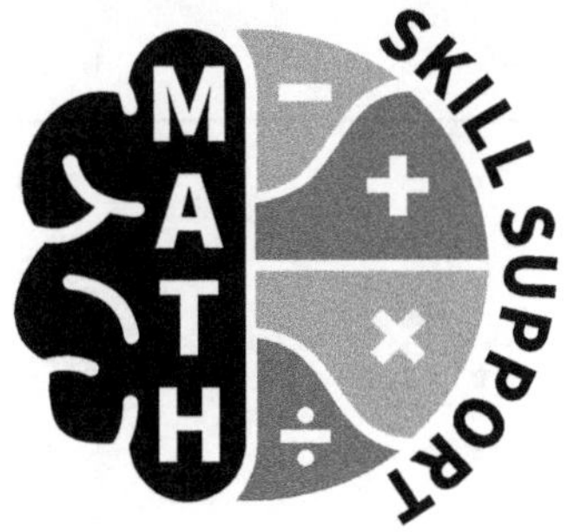

Name: ____________________ Date: ____________________

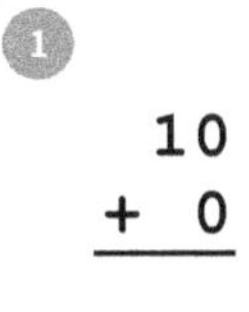

1. 10 + 0
2. 10 + 8
3. 5 + 5
4. 9 + 4
5. 10 + 4
6. 0 + 9
7. 1 + 3
8. 7 + 1
9. 14 + 3
10. 12 + 3
11. 6 + 5
12. 0 + 5
13. 1 + 6
14. 2 + 9
15. 2 + 3
16. 5 + 15
17. 4 + 6
18. 14 + 5
19. 1 + 2
20. 5 + 6
21. 5 + 14
22. 15 + 3
23. 11 + 3
24. 11 + 0
25. 10 + 5
26. 3 + 11
27. 15 + 5
28. 3 + 10
29. 4 + 7
30. 2 + 8
31. 15 + 4
32. 0 + 13
33. 2 + 4
34. 10 + 3
35. 18 + 0
36. 10 + 6
37. 3 + 12
38. 9 + 5
39. 7 + 4
40. 16 + 4
41. 0 + 17
42. 8 + 9
43. 9 + 6
44. 4 + 9
45. 9 + 11
46. 13 + 5
47. 0 + 3
48. 3 + 13
49. 8 + 5
50. 7 + 7
51. 2 + 17
52. 6 + 9
53. 6 + 3
54. 6 + 11
55. 5 + 0
56. 4 + 4
57. 9 + 10
58. 9 + 1
59. 14 + 0
60. 10 + 10

MATH SKILL SUPPORT

Name: ____________________ Date: ____________

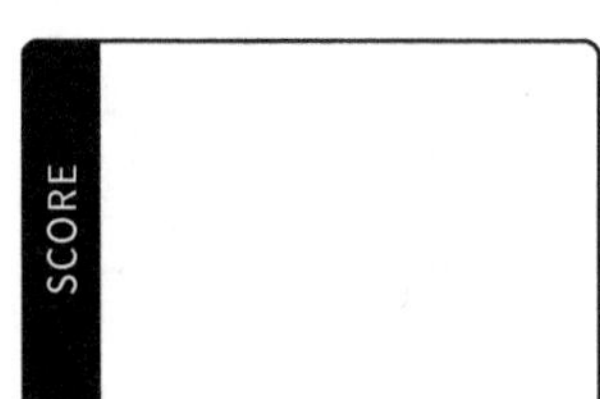
SCORE

1. 10 + 8	2. 7 + 13	3. 12 + 6	4. 4 + 6

5. 2 + 5	6. 2 + 18	7. 4 + 11	8. 8 + 6	9. 1 + 7	10. 17 + 1	11. 4 + 2	12. 14 + 2
13. 6 + 12	14. 5 + 12	15. 2 + 11	16. 3 + 6	17. 11 + 7	18. 13 + 4	19. 4 + 5	20. 16 + 4
21. 9 + 4	22. 4 + 15	23. 1 + 0	24. 12 + 4	25. 5 + 3	26. 3 + 8	27. 1 + 11	28. 10 + 5
29. 13 + 0	30. 0 + 17	31. 6 + 11	32. 5 + 8	33. 4 + 8	34. 2 + 3	35. 9 + 2	36. 18 + 1
37. 8 + 12	38. 1 + 6	39. 3 + 0	40. 5 + 9	41. 14 + 0	42. 13 + 1	43. 9 + 9	44. 11 + 6
45. 4 + 1	46. 3 + 13	47. 13 + 2	48. 15 + 4	49. 5 + 4	50. 20 + 0	51. 3 + 14	52. 11 + 4
53. 6 + 3	54. 1 + 4	55. 3 + 11	56. 9 + 0	57. 7 + 10	58. 6 + 4	59. 2 + 4	60. 5 + 5

MATH SKILL SUPPORT

Name: ______________________ Date: ______________

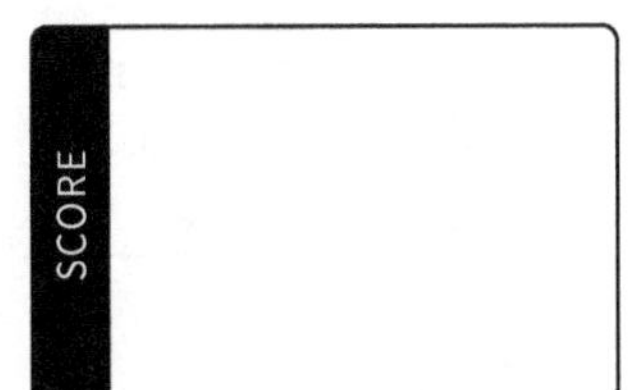

ADDING UP TO 20, SET H

				1 14 + 6	**2** 0 + 8	**3** 7 + 3	**4** 9 + 3
5 3 +10	**6** 4 + 4	**7** 9 + 6	**8** 10 + 2	**9** 11 + 2	**10** 3 + 3	**11** 8 + 4	**12** 9 + 7
13 0 +16	**14** 11 + 3	**15** 0 + 1	**16** 6 +13	**17** 19 + 0	**18** 6 + 2	**19** 9 +11	**20** 2 + 8
21 5 +11	**22** 6 + 7	**23** 2 +13	**24** 0 + 9	**25** 0 +18	**26** 16 + 2	**27** 5 + 1	**28** 8 + 3
29 0 +19	**30** 3 + 1	**31** 10 + 6	**32** 16 + 1	**33** 15 + 3	**34** 7 +11	**35** 13 + 5	**36** 19 + 1
37 6 + 5	**38** 6 +14	**39** 9 +10	**40** 8 + 7	**41** 16 + 3	**42** 0 + 3	**43** 1 +12	**44** 1 + 2
45 4 +10	**46** 1 + 8	**47** 8 +10	**48** 18 + 2	**49** 17 + 3	**50** 2 + 6	**51** 0 +20	**52** 1 +19
53 6 + 8	**54** 12 + 1	**55** 12 + 8	**56** 5 +10	**57** 10 + 3	**58** 3 + 9	**59** 14 + 4	**60** 0 + 7

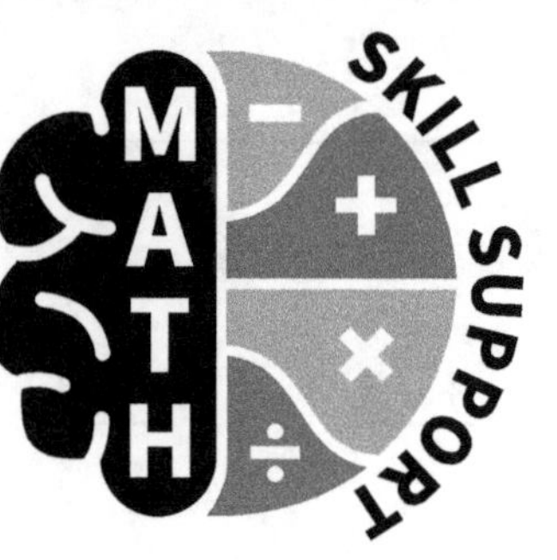

Name: ____________________ Date: ____________

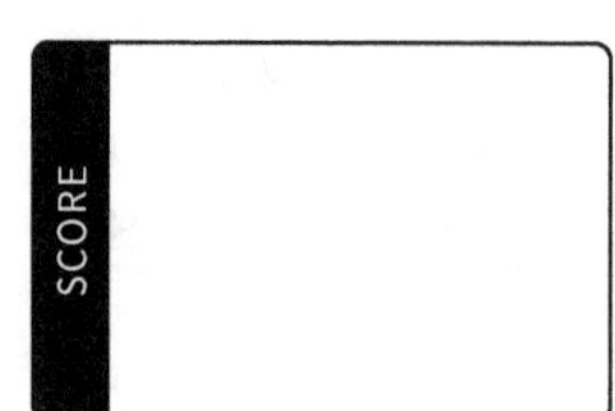

				(1) 10 + 10	(2) 11 + 0	(3) 2 + 10	(4) 13 + 7
(5) 7 + 8	(6) 1 + 13	(7) 2 + 7	(8) 9 + 1	(9) 3 + 15	(10) 8 + 9	(11) 6 + 1	(12) 4 + 16
(13) 10 + 9	(14) 17 + 0	(15) 7 + 6	(16) 5 + 0	(17) 10 + 1	(18) 0 + 12	(19) 9 + 5	(20) 12 + 2
(21) 3 + 17	(22) 0 + 14	(23) 6 + 9	(24) 0 + 2	(25) 13 + 3	(26) 0 + 11	(27) 3 + 12	(28) 8 + 5
(29) 12 + 5	(30) 7 + 2	(31) 2 + 16	(32) 18 + 0	(33) 15 + 0	(34) 1 + 15	(35) 2 + 2	(36) 7 + 5
(37) 12 + 3	(38) 11 + 1	(39) 1 + 5	(40) 0 + 6	(41) 15 + 2	(42) 4 + 12	(43) 4 + 14	(44) 0 + 0
(45) 1 + 17	(46) 2 + 9	(47) 1 + 3	(48) 10 + 0	(49) 2 + 0	(50) 5 + 7	(51) 1 + 1	(52) 5 + 14
(53) 6 + 0	(54) 1 + 14	(55) 7 + 4	(56) 11 + 5	(57) 10 + 4	(58) 2 + 1	(59) 3 + 4	(60) 4 + 0

Name: ______________________ Date: ______________

DAY 21

ADDING UP TO 20, SET J

1) 8 + 0	2) 4 + 16	3) 3 + 15	4) 0 + 5

5) 2 + 6	6) 0 + 0	7) 0 + 11	8) 3 + 12	9) 5 + 5	10) 1 + 13	11) 5 + 4	12) 1 + 8
13) 9 + 0	14) 0 + 15	15) 11 + 2	16) 7 + 13	17) 5 + 8	18) 20 + 0	19) 4 + 10	20) 0 + 13
21) 12 + 3	22) 11 + 0	23) 7 + 7	24) 10 + 8	25) 8 + 5	26) 12 + 6	27) 0 + 12	28) 5 + 10
29) 2 + 2	30) 7 + 4	31) 10 + 5	32) 6 + 0	33) 2 + 0	34) 2 + 3	35) 0 + 18	36) 4 + 11
37) 14 + 4	38) 14 + 0	39) 9 + 3	40) 15 + 4	41) 6 + 11	42) 11 + 7	43) 7 + 1	44) 16 + 2
45) 3 + 6	46) 2 + 12	47) 7 + 3	48) 17 + 3	49) 6 + 4	50) 4 + 0	51) 11 + 6	52) 1 + 9
53) 4 + 13	54) 15 + 2	55) 8 + 2	56) 19 + 1	57) 18 + 1	58) 13 + 3	59) 0 + 7	60) 3 + 10

MATH SKILL SUPPORT

Name: ______________________ Date: ______________

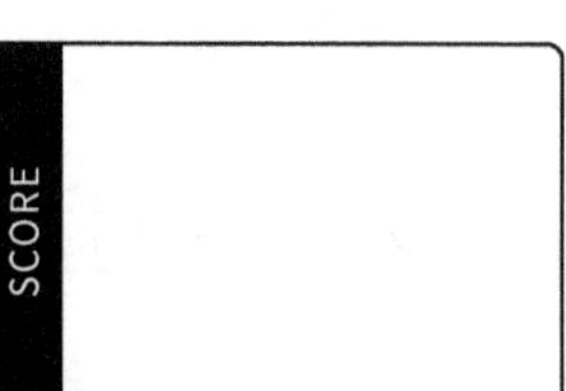
SCORE

				1. 0 + 8	2. 1 + 19	3. 8 + 11	4. 2 + 8
5. 0 + 14	6. 7 + 2	7. 3 + 11	8. 6 + 10	9. 0 + 3	10. 5 + 15	11. 4 + 12	12. 1 + 4
13. 11 + 3	14. 13 + 7	15. 2 + 14	16. 0 + 10	17. 18 + 2	18. 10 + 6	19. 9 + 7	20. 6 + 14
21. 1 + 2	22. 11 + 5	23. 2 + 1	24. 8 + 3	25. 1 + 14	26. 11 + 8	27. 7 + 0	28. 0 + 16
29. 12 + 2	30. 1 + 5	31. 6 + 13	32. 1 + 1	33. 0 + 9	34. 5 + 0	35. 4 + 5	36. 0 + 19
37. 8 + 6	38. 9 + 6	39. 16 + 1	40. 3 + 0	41. 3 + 5	42. 9 + 4	43. 3 + 17	44. 3 + 1
45. 8 + 4	46. 10 + 2	47. 6 + 12	48. 3 + 3	49. 0 + 4	50. 6 + 8	51. 13 + 4	52. 12 + 0
53. 5 + 7	54. 10 + 9	55. 12 + 5	56. 4 + 8	57. 17 + 2	58. 2 + 13	59. 10 + 4	60. 5 + 9

Name: ____________________ Date: ____________

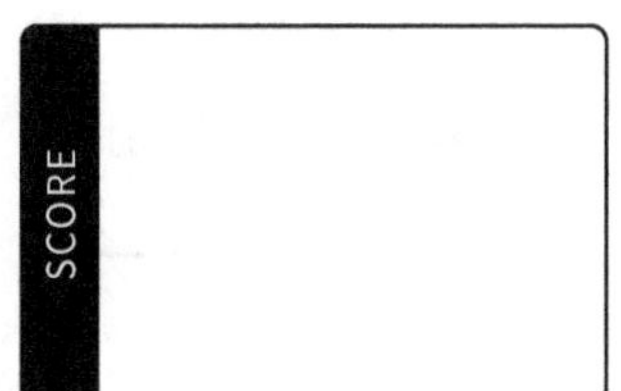

				1. 4 + 9	2. 2 + 7	3. 17 + 0	4. 9 +11
5. 12 + 1	6. 1 +17	7. 7 + 6	8. 11 + 9	9. 16 + 4	10. 3 + 2	11. 1 +10	12. 16 + 3
13. 0 +17	14. 10 + 3	15. 4 + 7	16. 5 +13	17. 2 +15	18. 9 + 1	19. 5 + 1	20. 6 + 6
21. 4 + 4	22. 8 + 9	23. 1 + 7	24. 10 +10	25. 6 + 7	26. 7 +11	27. 1 +18	28. 12 + 4
29. 3 + 9	30. 2 + 9	31. 5 + 3	32. 4 +15	33. 4 + 2	34. 4 + 6	35. 4 + 3	36. 16 + 0
37. 9 + 5	38. 5 +14	39. 15 + 1	40. 1 +16	41. 3 +14	42. 0 + 2	43. 1 +12	44. 10 + 7
45. 14 + 2	46. 5 + 2	47. 11 + 1	48. 7 +12	49. 8 + 8	50. 14 + 3	51. 6 + 1	52. 2 +11
53. 3 +13	54. 6 + 9	55. 5 +11	56. 14 + 1	57. 18 + 0	58. 7 + 9	59. 11 + 4	60. 15 + 3

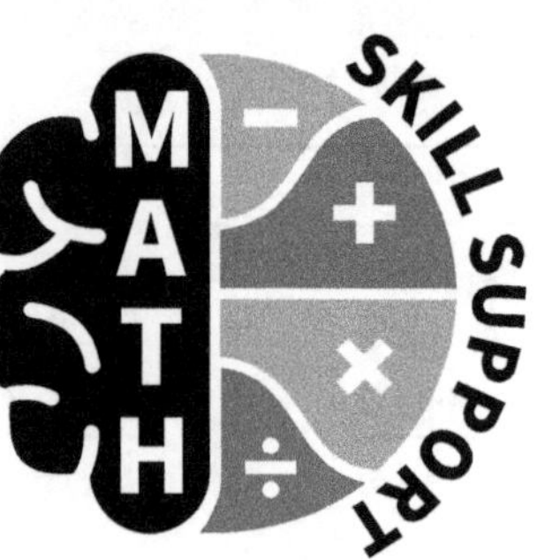

Name: ____________________ Date: ____________

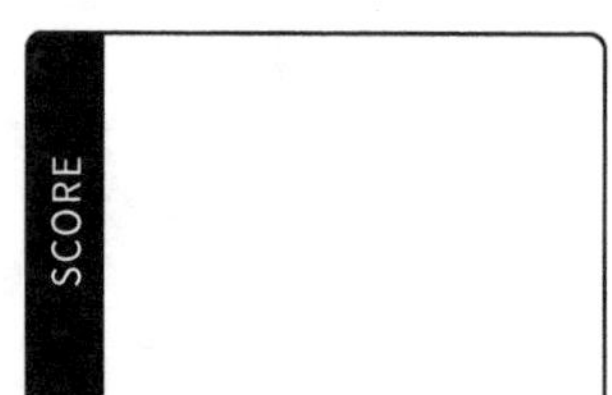

				1. 6 + 13	2. 14 + 2	3. 9 + 10	4. 2 + 9
5. 2 + 8	6. 3 + 5	7. 0 + 19	8. 8 + 10	9. 6 + 8	10. 6 + 0	11. 6 + 1	12. 12 + 5
13. 4 + 12	14. 4 + 0	15. 7 + 10	16. 12 + 2	17. 0 + 13	18. 3 + 7	19. 1 + 2	20. 5 + 7
21. 7 + 6	22. 1 + 6	23. 6 + 12	24. 5 + 1	25. 5 + 8	26. 11 + 1	27. 7 + 13	28. 8 + 5
29. 4 + 13	30. 10 + 8	31. 14 + 0	32. 15 + 0	33. 9 + 0	34. 18 + 0	35. 0 + 9	36. 12 + 6
37. 5 + 11	38. 12 + 4	39. 14 + 4	40. 14 + 6	41. 8 + 1	42. 11 + 8	43. 13 + 3	44. 0 + 6
45. 15 + 3	46. 7 + 3	47. 6 + 2	48. 15 + 4	49. 0 + 1	50. 4 + 4	51. 2 + 17	52. 9 + 4
53. 6 + 5	54. 1 + 16	55. 3 + 11	56. 9 + 2	57. 8 + 12	58. 7 + 0	59. 16 + 3	60. 4 + 10

MATH SKILL SUPPORT

Name: ____________________ Date: ____________

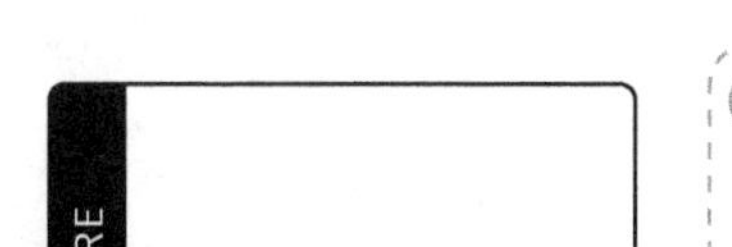

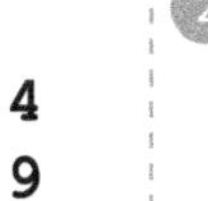

SCORE

DAY 25

ADDING UP TO 20, SET N

				1) 4 + 9	2) 9 + 1	3) 3 + 15	4) 15 + 1
5) 19 + 1	6) 13 + 7	7) 10 + 4	8) 7 + 9	9) 7 + 8	10) 3 + 14	11) 4 + 6	12) 11 + 3
13) 0 + 8	14) 6 + 3	15) 17 + 0	16) 16 + 0	17) 1 + 19	18) 0 + 2	19) 15 + 2	20) 0 + 14
21) 11 + 5	22) 4 + 2	23) 2 + 2	24) 10 + 5	25) 1 + 9	26) 10 + 2	27) 9 + 5	28) 5 + 13
29) 6 + 7	30) 1 + 7	31) 7 + 12	32) 3 + 9	33) 17 + 1	34) 12 + 7	35) 11 + 7	36) 2 + 3
37) 0 + 0	38) 5 + 3	39) 0 + 5	40) 2 + 12	41) 9 + 8	42) 6 + 6	43) 3 + 0	44) 12 + 8
45) 12 + 1	46) 8 + 0	47) 10 + 10	48) 13 + 0	49) 8 + 7	50) 11 + 6	51) 10 + 9	52) 1 + 13
53) 3 + 6	54) 4 + 1	55) 0 + 18	56) 6 + 4	57) 0 + 4	58) 17 + 3	59) 5 + 6	60) 2 + 16

MATH SKILL SUPPORT

Name: ______________________ Date: ______________

SCORE

1. 16 + 1
2. 12 + 3
3. 2 + 0
4. 0 + 15
5. 0 + 16
6. 0 + 12
7. 6 + 10
8. 9 + 6
9. 2 + 11
10. 4 + 7
11. 18 + 2
12. 5 + 14
13. 14 + 5
14. 1 + 3
15. 13 + 4
16. 5 + 0
17. 9 + 11
18. 8 + 3
19. 11 + 9
20. 3 + 17
21. 4 + 11
22. 10 + 6
23. 13 + 6
24. 9 + 3
25. 13 + 5
26. 5 + 2
27. 7 + 11
28. 4 + 15
29. 1 + 0
30. 1 + 17
31. 5 + 15
32. 8 + 2
33. 8 + 8
34. 4 + 16
35. 10 + 1
36. 8 + 4
37. 4 + 8
38. 14 + 3
39. 3 + 13
40. 11 + 4
41. 0 + 11
42. 3 + 16
43. 2 + 13
44. 10 + 0
45. 15 + 5
46. 19 + 0
47. 0 + 10
48. 6 + 9
49. 18 + 1
50. 0 + 7
51. 3 + 10
52. 1 + 15
53. 4 + 5
54. 0 + 17
55. 11 + 0
56. 2 + 6
57. 9 + 7
58. 10 + 7
59. 5 + 10
60. 2 + 4

MATH SKILL SUPPORT

Name: ____________________ Date: ____________________

SCORE

				1. 6 + 4	2. 6 + 3	3. 2 + 8	4. 6 + 11
5. 3 + 7	6. 10 + 9	7. 10 + 4	8. 15 + 4	9. 4 + 16	10. 9 + 3	11. 12 + 7	12. 2 + 10
13. 5 + 8	14. 3 + 8	15. 13 + 0	16. 18 + 1	17. 1 + 19	18. 3 + 12	19. 8 + 0	20. 12 + 5
21. 13 + 1	22. 13 + 7	23. 4 + 6	24. 0 + 14	25. 2 + 7	26. 2 + 5	27. 9 + 5	28. 7 + 0
29. 16 + 3	30. 10 + 10	31. 0 + 13	32. 4 + 8	33. 9 + 0	34. 16 + 1	35. 2 + 6	36. 10 + 2
37. 3 + 14	38. 1 + 0	39. 8 + 12	40. 7 + 10	41. 12 + 1	42. 4 + 10	43. 8 + 6	44. 11 + 5
45. 3 + 2	46. 8 + 7	47. 1 + 7	48. 8 + 1	49. 9 + 6	50. 11 + 3	51. 8 + 5	52. 3 + 13
53. 14 + 2	54. 16 + 2	55. 6 + 14	56. 3 + 5	57. 2 + 3	58. 11 + 8	59. 6 + 1	60. 0 + 19

Name: ______________________ Date: ______________

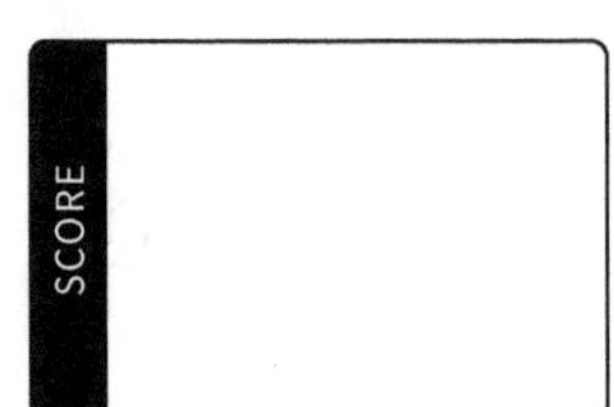

1	2	3	4
6 + 2	10 + 8	14 + 6	3 + 15

5	6	7	8	9	10	11	12
5 + 14	10 + 1	2 + 16	13 + 2	1 + 3	1 + 1	12 + 8	6 + 9

13	14	15	16	17	18	19	20
7 + 1	12 + 0	0 + 4	11 + 4	4 + 13	6 + 8	11 + 9	10 + 7

21	22	23	24	25	26	27	28
2 + 17	0 + 7	17 + 2	15 + 3	5 + 13	9 + 7	7 + 3	5 + 10

29	30	31	32	33	34	35	36
2 + 18	19 + 0	2 + 14	19 + 1	4 + 12	0 + 8	5 + 5	12 + 3

37	38	39	40	41	42	43	44
10 + 5	4 + 0	17 + 0	1 + 17	0 + 1	1 + 18	9 + 11	8 + 9

45	46	47	48	49	50	51	52
6 + 0	6 + 6	17 + 1	3 + 4	1 + 8	0 + 6	2 + 12	6 + 13

53	54	55	56	57	58	59	60
20 + 0	2 + 15	17 + 3	14 + 4	5 + 1	0 + 5	9 + 2	1 + 9

Name: ____________________ Date: ____________

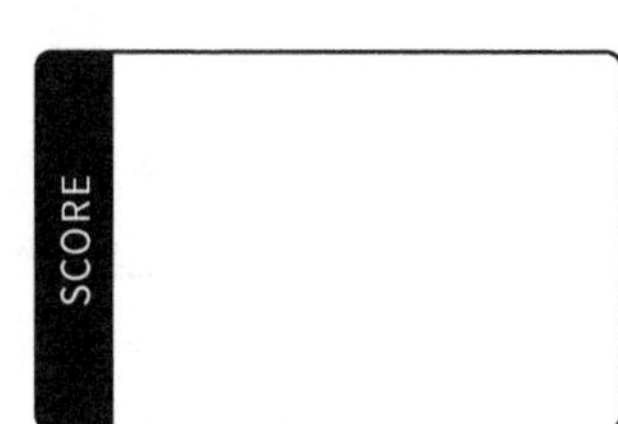

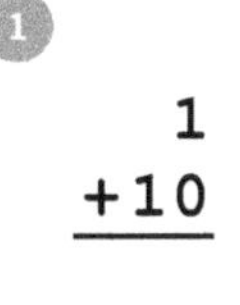

				(1) 1 +10	(2) 0 +12	(3) 4 + 3	(4) 8 +10
(5) 0 +18	(6) 7 + 5	(7) 4 + 5	(8) 14 + 1	(9) 10 + 0	(10) 5 + 4	(11) 3 + 6	(12) 1 +13
(13) 8 + 8	(14) 18 + 2	(15) 0 +11	(16) 2 + 2	(17) 6 + 7	(18) 7 + 6	(19) 2 + 9	(20) 12 + 6
(21) 2 + 0	(22) 2 +11	(23) 14 + 0	(24) 8 + 2	(25) 3 +17	(26) 11 + 7	(27) 1 + 5	(28) 13 + 3
(29) 15 + 2	(30) 5 + 3	(31) 5 + 6	(32) 1 +16	(33) 7 +12	(34) 8 + 4	(35) 3 +10	(36) 3 +11
(37) 0 + 9	(38) 2 + 4	(39) 0 + 2	(40) 13 + 5	(41) 1 + 6	(42) 4 +11	(43) 15 + 1	(44) 5 +11
(45) 16 + 0	(46) 3 + 3	(47) 4 +15	(48) 0 +10	(49) 0 +16	(50) 12 + 2	(51) 18 + 0	(52) 6 + 5
(53) 0 +15	(54) 0 + 3	(55) 1 + 4	(56) 6 +10	(57) 13 + 4	(58) 14 + 5	(59) 1 +11	(60) 7 + 7

MATH SKILL SUPPORT

Name: ____________________ Date: ____________________

SCORE

				1. 8 + 11	2. 16 + 4	3. 4 + 2	4. 6 + 2
5. 12 + 8	6. 2 + 1	7. 6 + 5	8. 7 + 11	9. 8 + 2	10. 5 + 13	11. 11 + 2	12. 4 + 15
13. 5 + 8	14. 1 + 12	15. 3 + 6	16. 6 + 8	17. 20 + 0	18. 14 + 3	19. 3 + 14	20. 0 + 3
21. 1 + 10	22. 2 + 11	23. 16 + 0	24. 7 + 1	25. 4 + 0	26. 1 + 9	27. 3 + 7	28. 3 + 5
29. 1 + 11	30. 9 + 5	31. 3 + 0	32. 11 + 0	33. 6 + 7	34. 18 + 2	35. 3 + 1	36. 3 + 4
37. 0 + 0	38. 10 + 2	39. 13 + 6	40. 0 + 18	41. 2 + 14	42. 3 + 3	43. 5 + 11	44. 5 + 15
45. 5 + 6	46. 19 + 1	47. 4 + 11	48. 12 + 2	49. 1 + 5	50. 6 + 4	51. 13 + 1	52. 12 + 6
53. 8 + 7	54. 8 + 9	55. 1 + 3	56. 6 + 12	57. 5 + 12	58. 9 + 9	59. 1 + 2	60. 10 + 3

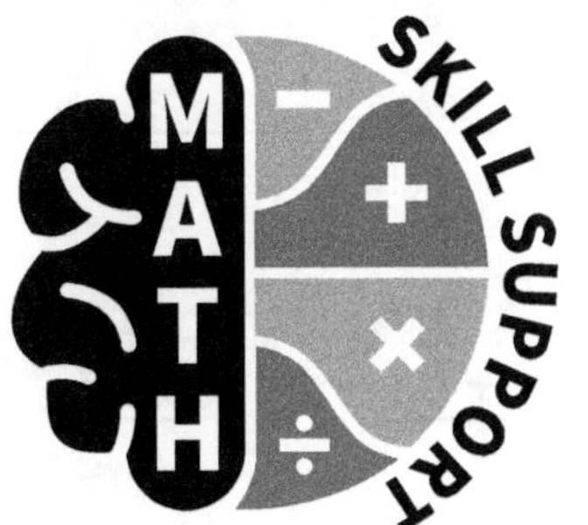

Name: ______________________ Date: ______________________

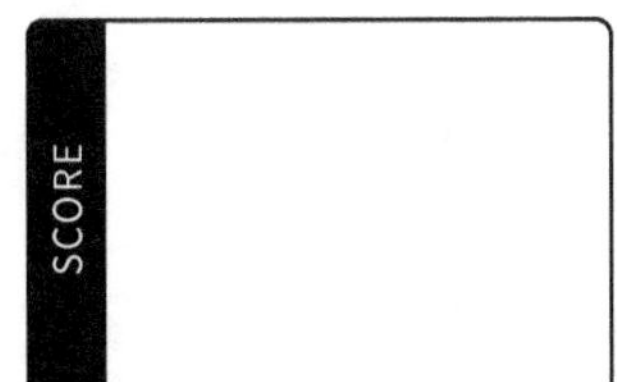

#	Problem
1	0 + 1
2	0 + 12
3	7 + 10
4	9 + 6
5	1 + 8
6	3 + 13
7	10 + 10
8	5 + 14
9	13 + 5
10	3 + 10
11	2 + 5
12	2 + 13
13	10 + 0
14	12 + 3
15	1 + 17
16	1 + 16
17	0 + 13
18	5 + 3
19	5 + 7
20	12 + 7
21	14 + 1
22	6 + 11
23	7 + 0
24	9 + 2
25	12 + 1
26	4 + 13
27	0 + 11
28	1 + 4
29	2 + 16
30	0 + 17
31	8 + 1
32	9 + 8
33	4 + 1
34	16 + 1
35	14 + 4
36	8 + 6
37	2 + 7
38	10 + 9
39	16 + 3
40	4 + 14
41	1 + 1
42	4 + 10
43	9 + 1
44	15 + 3
45	1 + 13
46	5 + 1
47	10 + 7
48	0 + 10
49	14 + 2
50	7 + 6
51	17 + 0
52	11 + 5
53	11 + 1
54	16 + 2
55	12 + 0
56	4 + 6
57	0 + 8
58	0 + 15
59	5 + 4
60	15 + 4

DAY 32

ADDING UP TO 20, SET U

MATH SKILL SUPPORT

Name: ____________________ Date: ____________

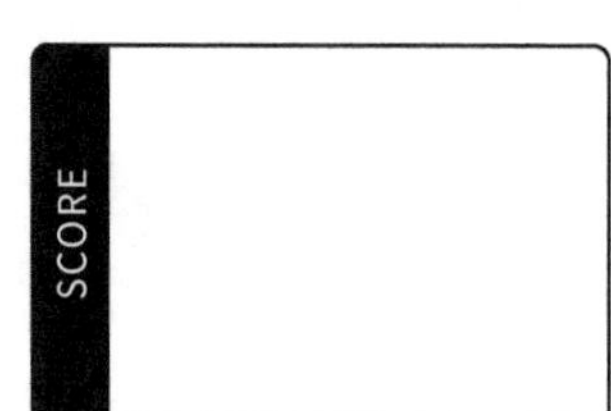

				(1) 14 + 6	(2) 3 + 17	(3) 17 + 2	(4) 13 + 7
(5) 2 + 12	(6) 9 + 0	(7) 2 + 6	(8) 10 + 5	(9) 6 + 6	(10) 15 + 1	(11) 10 + 6	(12) 4 + 9
(13) 13 + 4	(14) 1 + 0	(15) 2 + 17	(16) 4 + 4	(17) 17 + 1	(18) 6 + 13	(19) 7 + 5	(20) 1 + 14
(21) 7 + 3	(22) 7 + 4	(23) 4 + 5	(24) 15 + 5	(25) 0 + 4	(26) 19 + 0	(27) 3 + 9	(28) 13 + 0
(29) 11 + 6	(30) 7 + 7	(31) 0 + 9	(32) 7 + 12	(33) 0 + 20	(34) 3 + 15	(35) 1 + 7	(36) 1 + 19
(37) 17 + 3	(38) 3 + 16	(39) 2 + 0	(40) 11 + 9	(41) 2 + 2	(42) 5 + 10	(43) 8 + 3	(44) 1 + 6
(45) 8 + 0	(46) 10 + 1	(47) 10 + 4	(48) 10 + 8	(49) 4 + 7	(50) 6 + 10	(51) 13 + 3	(52) 6 + 14
(53) 1 + 15	(54) 4 + 12	(55) 7 + 9	(56) 11 + 4	(57) 8 + 10	(58) 8 + 4	(59) 0 + 6	(60) 5 + 0

Name: ____________________ Date: ____________

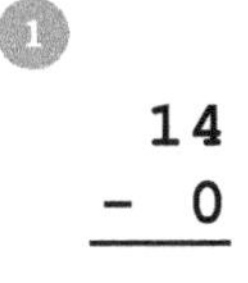
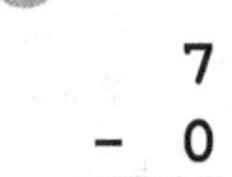

DAY 33

SUBTRACTING 0, SET A

				(1) 14 − 0	(2) 10 − 0	(3) 17 − 0	(4) 7 − 0
(5) 0 − 0	(6) 5 − 0	(7) 6 − 0	(8) 13 − 0	(9) 20 − 0	(10) 4 − 0	(11) 8 − 0	(12) 12 − 0
(13) 18 − 0	(14) 11 − 0	(15) 19 − 0	(16) 9 − 0	(17) 16 − 0	(18) 1 − 0	(19) 15 − 0	(20) 2 − 0
(21) 3 − 0	(22) 11 − 0	(23) 7 − 0	(24) 0 − 0	(25) 18 − 0	(26) 14 − 0	(27) 6 − 0	(28) 16 − 0
(29) 10 − 0	(30) 3 − 0	(31) 13 − 0	(32) 1 − 0	(33) 9 − 0	(34) 17 − 0	(35) 4 − 0	(36) 19 − 0
(37) 8 − 0	(38) 15 − 0	(39) 20 − 0	(40) 5 − 0	(41) 2 − 0	(42) 12 − 0	(43) 1 − 0	(44) 10 − 0
(45) 20 − 0	(46) 18 − 0	(47) 4 − 0	(48) 19 − 0	(49) 7 − 0	(50) 0 − 0	(51) 6 − 0	(52) 3 − 0
(53) 5 − 0	(54) 14 − 0	(55) 17 − 0	(56) 9 − 0	(57) 2 − 0	(58) 16 − 0	(59) 8 − 0	(60) 12 − 0

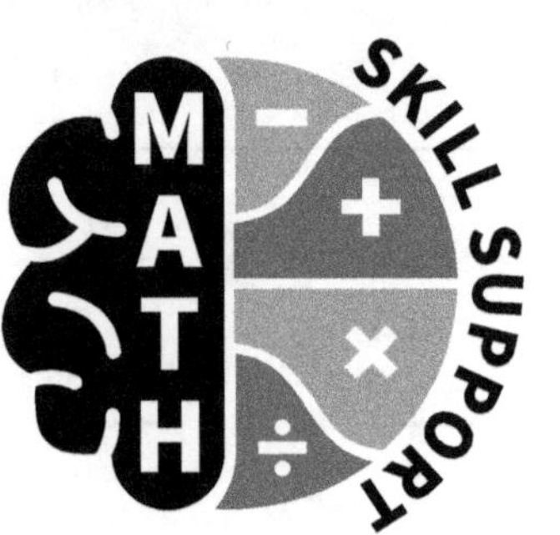

Name: ____________________ Date: ____________

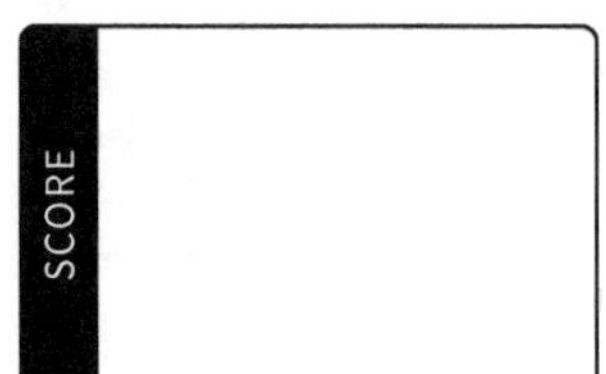

				1. 2 − 0	2. 13 − 0	3. 11 − 0	4. 5 − 0
5. 16 − 0	6. 9 − 0	7. 0 − 0	8. 17 − 0	9. 3 − 0	10. 12 − 0	11. 19 − 0	12. 1 − 0
13. 6 − 0	14. 7 − 0	15. 14 − 0	16. 8 − 0	17. 10 − 0	18. 4 − 0	19. 20 − 0	20. 18 − 0
21. 15 − 0	22. 8 − 0	23. 13 − 0	24. 16 − 0	25. 4 − 0	26. 19 − 0	27. 14 − 0	28. 10 − 0
29. 7 − 0	30. 15 − 0	31. 1 − 0	32. 12 − 0	33. 5 − 0	34. 6 − 0	35. 2 − 0	36. 3 − 0
37. 11 − 0	38. 0 − 0	39. 20 − 0	40. 17 − 0	41. 18 − 0	42. 9 − 0	43. 15 − 0	44. 13 − 0
45. 5 − 0	46. 0 − 0	47. 18 − 0	48. 8 − 0	49. 7 − 0	50. 1 − 0	51. 14 − 0	52. 16 − 0
53. 17 − 0	54. 19 − 0	55. 11 − 0	56. 3 − 0	57. 20 − 0	58. 12 − 0	59. 6 − 0	60. 9 − 0

MATH SKILL SUPPORT

DAY 35

SUBTRACTING 1, SET A

Name: ____________________ Date: ____________

SCORE

				1) $10 - 1$	2) $20 - 1$	3) $1 - 0$	4) $6 - 1$
5) $2 - 1$	6) $5 - 1$	7) $7 - 1$	8) $15 - 1$	9) $18 - 1$	10) $19 - 1$	11) $13 - 1$	12) $4 - 1$
13) $14 - 1$	14) $17 - 1$	15) $12 - 1$	16) $16 - 1$	17) $1 - 1$	18) $9 - 1$	19) $11 - 1$	20) $3 - 1$
21) $8 - 1$	22) $12 - 1$	23) $16 - 1$	24) $9 - 1$	25) $4 - 1$	26) $1 - 0$	27) $15 - 1$	28) $14 - 1$
29) $2 - 1$	30) $13 - 1$	31) $11 - 1$	32) $8 - 1$	33) $18 - 1$	34) $6 - 1$	35) $1 - 1$	36) $17 - 1$
37) $20 - 1$	38) $19 - 1$	39) $10 - 1$	40) $5 - 1$	41) $3 - 1$	42) $7 - 1$	43) $9 - 1$	44) $2 - 1$
45) $16 - 1$	46) $8 - 1$	47) $5 - 1$	48) $14 - 1$	49) $11 - 1$	50) $18 - 1$	51) $4 - 1$	52) $7 - 1$
53) $15 - 1$	54) $13 - 1$	55) $10 - 1$	56) $12 - 1$	57) $17 - 1$	58) $1 - 0$	59) $19 - 1$	60) $3 - 1$

MATH SKILL SUPPORT

Name: ____________________ Date: ____________

				(1) 20 − 1	(2) 1 − 0	(3) 3 − 1	(4) 11 − 1
(5) 4 − 1	(6) 17 − 1	(7) 7 − 1	(8) 12 − 1	(9) 14 − 1	(10) 19 − 1	(11) 16 − 1	(12) 9 − 1
(13) 5 − 1	(14) 15 − 1	(15) 10 − 1	(16) 8 − 1	(17) 1 − 1	(18) 18 − 1	(19) 13 − 1	(20) 6 − 1
(21) 2 − 1	(22) 9 − 1	(23) 12 − 1	(24) 18 − 1	(25) 5 − 1	(26) 1 − 0	(27) 10 − 1	(28) 15 − 1
(29) 20 − 1	(30) 1 − 1	(31) 8 − 1	(32) 11 − 1	(33) 2 − 1	(34) 4 − 1	(35) 17 − 1	(36) 6 − 1
(37) 3 − 1	(38) 19 − 1	(39) 13 − 1	(40) 14 − 1	(41) 16 − 1	(42) 7 − 1	(43) 9 − 1	(44) 4 − 1
(45) 16 − 1	(46) 7 − 1	(47) 17 − 1	(48) 15 − 1	(49) 20 − 1	(50) 6 − 1	(51) 1 − 0	(52) 1 − 1
(53) 2 − 1	(54) 8 − 1	(55) 11 − 1	(56) 3 − 1	(57) 10 − 1	(58) 18 − 1	(59) 14 − 1	(60) 19 − 1

Name: ____________________ Date: ____________________

SCORE

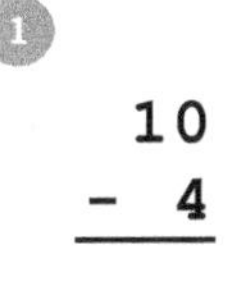

				1. 10 - 4	2. 13 - 10	3. 10 - 9	4. 19 - 10
5. 17 - 10	6. 10 - 2	7. 10 - 6	8. 10 - 5	9. 20 - 10	10. 10 - 1	11. 10 - 0	12. 15 - 10
13. 10 - 7	14. 11 - 10	15. 10 - 10	16. 10 - 3	17. 10 - 8	18. 14 - 10	19. 18 - 10	20. 16 - 10
21. 12 - 10	22. 10 - 6	23. 19 - 10	24. 10 - 8	25. 17 - 10	26. 11 - 10	27. 18 - 10	28. 10 - 7
29. 10 - 5	30. 12 - 10	31. 10 - 3	32. 16 - 10	33. 10 - 4	34. 10 - 0	35. 10 - 9	36. 15 - 10
37. 13 - 10	38. 10 - 10	39. 20 - 10	40. 10 - 1	41. 10 - 2	42. 14 - 10	43. 19 - 10	44. 15 - 10
45. 10 - 6	46. 20 - 10	47. 10 - 3	48. 16 - 10	49. 10 - 1	50. 10 - 10	51. 10 - 2	52. 17 - 10
53. 14 - 10	54. 10 - 8	55. 11 - 10	56. 10 - 7	57. 10 - 5	58. 18 - 10	59. 10 - 4	60. 12 - 10

MATH SKILL SUPPORT

Name: ____________________ Date: ____________________

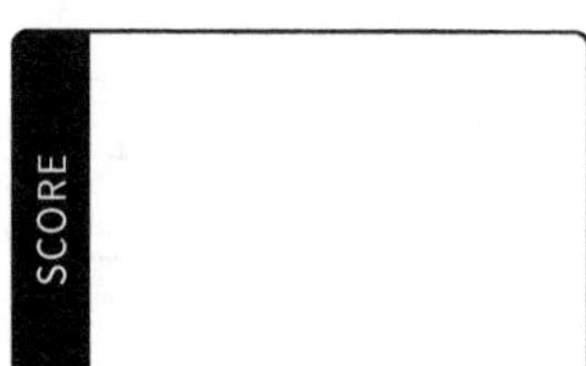

1. $10 - 10$	2. $10 - 8$	3. $16 - 10$	4. $19 - 10$

5. $10 - 2$	6. $17 - 10$	7. $10 - 0$	8. $10 - 1$	9. $13 - 10$	10. $10 - 3$	11. $12 - 10$	12. $15 - 10$
13. $11 - 10$	14. $10 - 9$	15. $10 - 5$	16. $10 - 4$	17. $14 - 10$	18. $20 - 10$	19. $10 - 7$	20. $18 - 10$
21. $10 - 6$	22. $10 - 5$	23. $16 - 10$	24. $20 - 10$	25. $10 - 7$	26. $10 - 1$	27. $19 - 10$	28. $18 - 10$
29. $12 - 10$	30. $11 - 10$	31. $10 - 3$	32. $10 - 2$	33. $13 - 10$	34. $17 - 10$	35. $10 - 0$	36. $15 - 10$
37. $10 - 4$	38. $10 - 10$	39. $10 - 6$	40. $10 - 9$	41. $14 - 10$	42. $10 - 8$	43. $14 - 10$	44. $17 - 10$
45. $10 - 2$	46. $20 - 10$	47. $13 - 10$	48. $10 - 8$	49. $11 - 10$	50. $10 - 3$	51. $18 - 10$	52. $15 - 10$
53. $10 - 5$	54. $10 - 4$	55. $10 - 9$	56. $12 - 10$	57. $10 - 7$	58. $10 - 0$	59. $10 - 6$	60. $19 - 10$

SUBTRACTING WITHIN 10, SET A

Name: ____________________ Date: ____________

#	Problem	#	Problem	#	Problem	#	Problem
1	$2 - 2$	2	$9 - 7$	3	$9 - 4$	4	$7 - 3$
5	$4 - 0$	6	$10 - 1$	7	$8 - 5$	8	$5 - 0$
9	$10 - 10$	10	$7 - 2$	11	$5 - 2$	12	$10 - 7$
13	$6 - 1$	14	$9 - 5$	15	$7 - 6$	16	$6 - 3$
17	$2 - 1$	18	$9 - 3$	19	$0 - 0$	20	$10 - 0$
21	$6 - 2$	22	$4 - 3$	23	$9 - 0$	24	$10 - 2$
25	$8 - 4$	26	$7 - 5$	27	$10 - 8$	28	$1 - 0$
29	$8 - 1$	30	$10 - 4$	31	$4 - 4$	32	$2 - 0$
33	$10 - 5$	34	$9 - 2$	35	$9 - 9$	36	$4 - 2$
37	$10 - 6$	38	$6 - 4$	39	$7 - 0$	40	$7 - 1$
41	$8 - 7$	42	$3 - 1$	43	$5 - 4$	44	$9 - 1$
45	$6 - 0$	46	$7 - 7$	47	$7 - 4$	48	$9 - 8$
49	$8 - 0$	50	$5 - 1$	51	$8 - 8$	52	$5 - 5$
53	$8 - 2$	54	$10 - 3$	55	$9 - 6$	56	$4 - 1$
57	$3 - 2$	58	$10 - 9$	59	$5 - 3$	60	$6 - 6$

MATH SKILL SUPPORT

Name: ______________________ Date: ______________

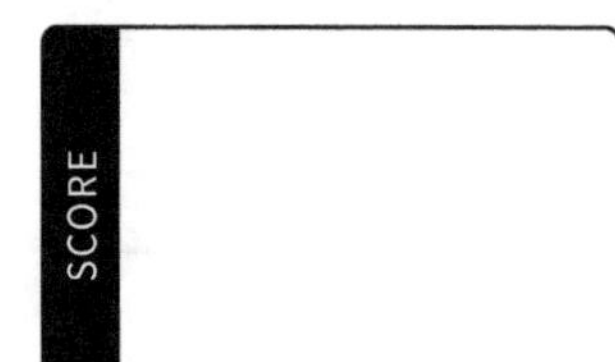

				(1) 7 − 7	(2) 2 − 2	(3) 10 − 7	(4) 4 − 1
(5) 6 − 3	(6) 5 − 1	(7) 8 − 0	(8) 5 − 0	(9) 8 − 1	(10) 10 − 2	(11) 4 − 0	(12) 8 − 7
(13) 4 − 2	(14) 7 − 0	(15) 9 − 5	(16) 7 − 6	(17) 5 − 5	(18) 8 − 8	(19) 9 − 1	(20) 6 − 2
(21) 3 − 3	(22) 9 − 3	(23) 3 − 2	(24) 7 − 5	(25) 6 − 1	(26) 8 − 5	(27) 5 − 2	(28) 10 − 9
(29) 10 − 4	(30) 6 − 4	(31) 10 − 1	(32) 1 − 1	(33) 8 − 4	(34) 7 − 1	(35) 3 − 0	(36) 8 − 6
(37) 8 − 3	(38) 9 − 9	(39) 10 − 5	(40) 9 − 8	(41) 6 − 5	(42) 9 − 6	(43) 9 − 7	(44) 7 − 4
(45) 9 − 4	(46) 5 − 3	(47) 3 − 1	(48) 10 − 6	(49) 7 − 3	(50) 4 − 4	(51) 7 − 2	(52) 4 − 3
(53) 10 − 10	(54) 0 − 0	(55) 6 − 0	(56) 10 − 3	(57) 6 − 6	(58) 10 − 8	(59) 10 − 0	(60) 9 − 0

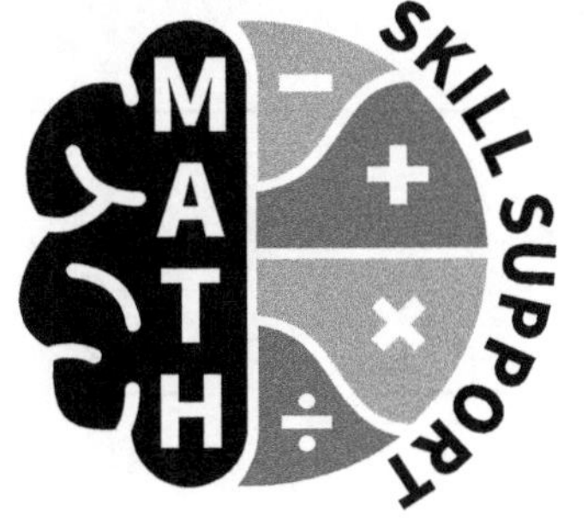

Name: ______________________ Date: ______________

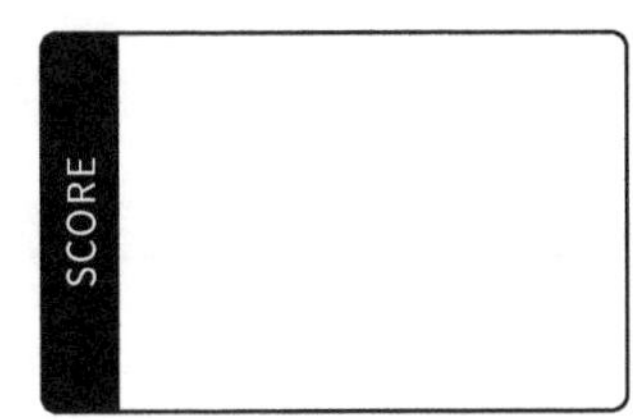

SUBTRACTING WITHIN 10, SET C

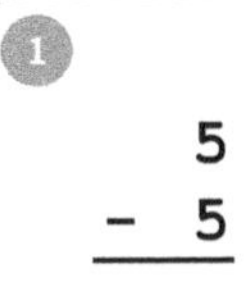

				1. $5 - 5$	2. $9 - 3$	3. $10 - 7$	4. $6 - 6$
5. $9 - 1$	6. $5 - 3$	7. $10 - 3$	8. $4 - 2$	9. $9 - 4$	10. $6 - 1$	11. $8 - 4$	12. $10 - 0$
13. $7 - 6$	14. $8 - 8$	15. $4 - 1$	16. $2 - 0$	17. $5 - 4$	18. $8 - 1$	19. $7 - 0$	20. $9 - 8$
21. $2 - 1$	22. $1 - 1$	23. $6 - 2$	24. $10 - 9$	25. $8 - 7$	26. $7 - 1$	27. $7 - 5$	28. $10 - 8$
29. $8 - 2$	30. $9 - 2$	31. $6 - 5$	32. $10 - 1$	33. $9 - 0$	34. $3 - 3$	35. $8 - 6$	36. $10 - 5$
37. $9 - 5$	38. $10 - 10$	39. $1 - 0$	40. $8 - 5$	41. $0 - 0$	42. $6 - 3$	43. $7 - 2$	44. $4 - 0$
45. $2 - 2$	46. $7 - 7$	47. $3 - 0$	48. $5 - 2$	49. $8 - 3$	50. $10 - 2$	51. $6 - 4$	52. $9 - 6$
53. $7 - 3$	54. $8 - 0$	55. $3 - 2$	56. $9 - 9$	57. $3 - 1$	58. $5 - 1$	59. $9 - 7$	60. $4 - 4$

MATH SKILL SUPPORT

Name: ____________________ Date: ____________________

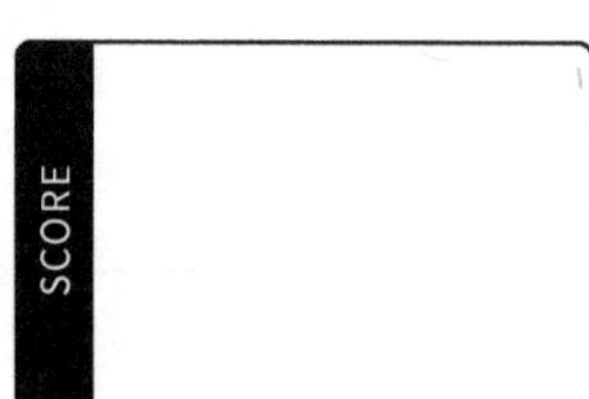
SCORE

				1. $7 - 0$	2. $9 - 2$	3. $8 - 1$	4. $6 - 3$
5. $1 - 0$	6. $7 - 2$	7. $5 - 4$	8. $3 - 2$	9. $9 - 7$	10. $10 - 6$	11. $10 - 3$	12. $10 - 4$
13. $7 - 3$	14. $5 - 3$	15. $4 - 2$	16. $10 - 9$	17. $6 - 6$	18. $6 - 4$	19. $6 - 5$	20. $3 - 0$
21. $4 - 3$	22. $8 - 0$	23. $4 - 4$	24. $4 - 1$	25. $2 - 0$	26. $9 - 8$	27. $5 - 2$	28. $10 - 5$
29. $9 - 5$	30. $7 - 4$	31. $9 - 0$	32. $10 - 0$	33. $9 - 9$	34. $6 - 2$	35. $6 - 0$	36. $8 - 8$
37. $8 - 3$	38. $2 - 1$	39. $1 - 1$	40. $8 - 4$	41. $7 - 7$	42. $7 - 6$	43. $10 - 1$	44. $4 - 0$
45. $7 - 1$	46. $10 - 8$	47. $9 - 6$	48. $6 - 1$	49. $8 - 7$	50. $8 - 6$	51. $3 - 3$	52. $10 - 10$
53. $5 - 1$	54. $5 - 0$	55. $7 - 5$	56. $5 - 5$	57. $10 - 7$	58. $8 - 2$	59. $9 - 4$	60. $9 - 3$

Name: ______________________ Date: ______________

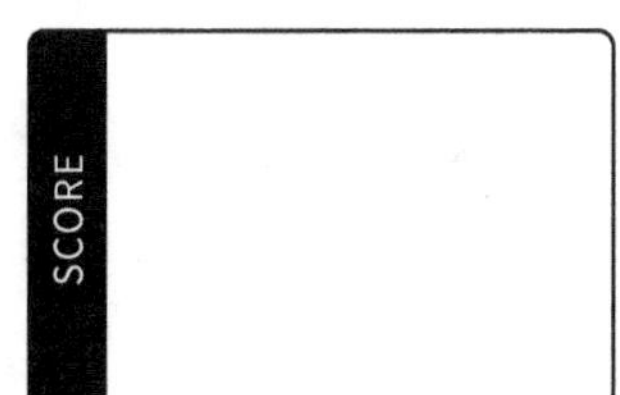

1	2	3	4
$10 - 8$	$8 - 1$	$9 - 7$	$2 - 0$

5	6	7	8	9	10	11	12
$7 - 6$	$3 - 1$	$4 - 0$	$5 - 2$	$6 - 0$	$7 - 3$	$6 - 5$	$3 - 3$

13	14	15	16	17	18	19	20
$5 - 5$	$10 - 5$	$5 - 3$	$10 - 7$	$7 - 4$	$3 - 2$	$4 - 4$	$6 - 4$

21	22	23	24	25	26	27	28
$10 - 6$	$4 - 2$	$8 - 6$	$6 - 6$	$2 - 2$	$9 - 4$	$8 - 5$	$6 - 3$

29	30	31	32	33	34	35	36
$1 - 0$	$10 - 2$	$10 - 3$	$0 - 0$	$9 - 6$	$5 - 1$	$3 - 0$	$9 - 8$

37	38	39	40	41	42	43	44
$7 - 1$	$10 - 4$	$7 - 0$	$1 - 1$	$10 - 0$	$10 - 1$	$8 - 4$	$9 - 2$

45	46	47	48	49	50	51	52
$2 - 1$	$4 - 3$	$5 - 0$	$5 - 4$	$9 - 9$	$9 - 3$	$8 - 7$	$7 - 2$

53	54	55	56	57	58	59	60
$8 - 2$	$7 - 5$	$6 - 2$	$10 - 9$	$8 - 3$	$7 - 7$	$8 - 8$	$10 - 10$

SUBTRACTING WITHIN 20, SET A

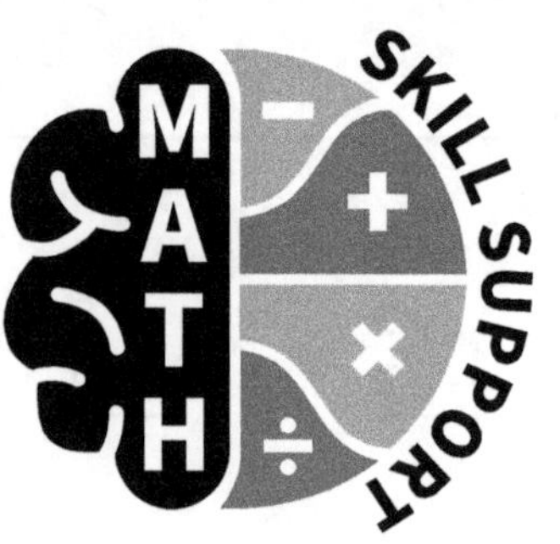

Name: ______________________ Date: ______________________

				1. 6 - 3	2. 4 - 3	3. 15 - 8	4. 9 - 2
5. 16 - 2	6. 18 - 15	7. 11 - 7	8. 20 - 5	9. 10 - 1	10. 6 - 0	11. 15 - 10	12. 9 - 8
13. 17 - 1	14. 20 - 0	15. 15 - 13	16. 12 - 11	17. 16 - 6	18. 19 - 8	19. 19 - 19	20. 10 - 10
21. 17 - 5	22. 11 - 0	23. 20 - 13	24. 5 - 2	25. 19 - 1	26. 1 - 1	27. 13 - 0	28. 18 - 8
29. 2 - 0	30. 17 - 0	31. 19 - 10	32. 9 - 1	33. 14 - 2	34. 17 - 8	35. 4 - 0	36. 19 - 3
37. 20 - 19	38. 19 - 15	39. 16 - 16	40. 19 - 11	41. 18 - 6	42. 19 - 5	43. 20 - 12	44. 9 - 0
45. 2 - 1	46. 18 - 0	47. 15 - 5	48. 17 - 13	49. 20 - 10	50. 14 - 5	51. 17 - 14	52. 18 - 13
53. 7 - 1	54. 4 - 2	55. 11 - 9	56. 12 - 2	57. 17 - 16	58. 14 - 0	59. 13 - 7	60. 20 - 11

DAY 45 SUBTRACTING WITHIN 20, SET B

MATH SKILL SUPPORT

Name: ______________________ Date: ______________

SCORE

				1. 13 − 13	2. 17 − 15	3. 14 − 10	4. 10 − 7
5. 17 − 7	6. 14 − 4	7. 14 − 1	8. 9 − 3	9. 13 − 3	10. 8 − 0	11. 7 − 2	12. 10 − 6
13. 19 − 18	14. 18 − 1	15. 18 − 16	16. 17 − 2	17. 19 − 12	18. 12 − 9	19. 20 − 14	20. 15 − 6
21. 8 − 1	22. 14 − 13	23. 19 − 2	24. 6 − 5	25. 11 − 8	26. 17 − 12	27. 16 − 15	28. 11 − 5
29. 7 − 0	30. 18 − 3	31. 16 − 5	32. 8 − 7	33. 11 − 2	34. 8 − 2	35. 10 − 2	36. 12 − 0
37. 10 − 0	38. 20 − 15	39. 14 − 7	40. 16 − 14	41. 15 − 14	42. 16 − 0	43. 14 − 3	44. 7 − 4
45. 16 − 12	46. 9 − 9	47. 6 − 4	48. 12 − 12	49. 20 − 6	50. 8 − 8	51. 15 − 2	52. 6 − 6
53. 15 − 12	54. 15 − 15	55. 13 − 9	56. 3 − 2	57. 8 − 6	58. 19 − 4	59. 10 − 3	60. 14 − 14

DAY 46

SUBTRACTING WITHIN 20, SET C

MATH SKILL SUPPORT

Name: ____________________ Date: ____________________

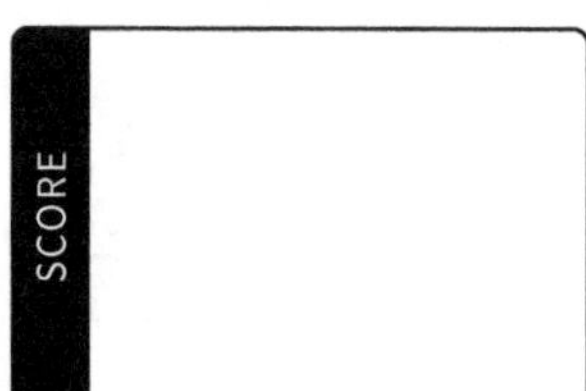

				1. 19 − 16	2. 11 − 4	3. 9 − 4	4. 15 − 4
5. 5 − 3	6. 18 − 7	7. 16 − 10	8. 16 − 7	9. 17 − 4	10. 12 − 10	11. 7 − 5	12. 14 − 8
13. 7 − 6	14. 8 − 5	15. 18 − 2	16. 19 − 17	17. 17 − 9	18. 13 − 1	19. 11 − 6	20. 18 − 12
21. 13 − 10	22. 17 − 3	23. 16 − 8	24. 16 − 13	25. 20 − 3	26. 3 − 0	27. 5 − 1	28. 16 − 9
29. 9 − 6	30. 19 − 13	31. 12 − 5	32. 12 − 3	33. 5 − 5	34. 4 − 1	35. 18 − 18	36. 15 − 0
37. 20 − 16	38. 5 − 0	39. 20 − 18	40. 16 − 3	41. 16 − 11	42. 12 − 8	43. 6 − 1	44. 8 − 4
45. 18 − 10	46. 20 − 4	47. 17 − 6	48. 6 − 2	49. 19 − 7	50. 10 − 8	51. 7 − 7	52. 18 − 11
53. 3 − 1	54. 16 − 1	55. 12 − 7	56. 20 − 1	57. 13 − 4	58. 7 − 3	59. 12 − 4	60. 2 − 2

SUBTRACTING WITHIN 20, SET D

MATH SKILL SUPPORT

Name: ____________________ Date: ____________________

SCORE

1. 19 − 10	2. 11 − 11	3. 15 − 15	4. 7 − 1

5. 7 − 4	6. 2 − 1	7. 12 − 0	8. 17 − 3	9. 4 − 3	10. 14 − 3	11. 18 − 12	12. 10 − 6
13. 12 − 10	14. 20 − 15	15. 6 − 3	16. 16 − 4	17. 16 − 9	18. 7 − 5	19. 15 − 12	20. 13 − 12
21. 15 − 14	22. 16 − 10	23. 12 − 2	24. 15 − 9	25. 12 − 12	26. 17 − 17	27. 6 − 4	28. 4 − 2
29. 17 − 13	30. 14 − 4	31. 5 − 3	32. 19 − 7	33. 10 − 3	34. 11 − 7	35. 20 − 1	36. 19 − 3
37. 9 − 2	38. 18 − 5	39. 10 − 7	40. 6 − 0	41. 15 − 7	42. 9 − 3	43. 6 − 5	44. 20 − 5
45. 18 − 13	46. 18 − 2	47. 14 − 2	48. 13 − 1	49. 11 − 0	50. 18 − 1	51. 17 − 4	52. 11 − 8
53. 13 − 9	54. 13 − 8	55. 13 − 6	56. 19 − 17	57. 14 − 12	58. 16 − 7	59. 14 − 5	60. 9 − 4

SUBTRACTING WITHIN 20, SET E

MATH SKILL SUPPORT

Name: ______________________ Date: ______________

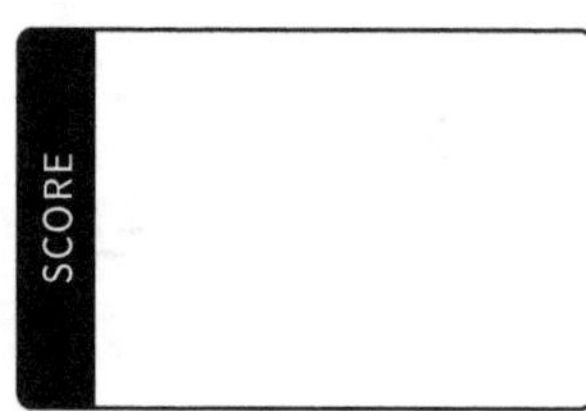

				1. 7 − 6	2. 8 − 4	3. 8 − 0	4. 15 − 1
5. 10 − 5	6. 19 − 2	7. 9 − 0	8. 11 − 9	9. 19 − 0	10. 15 − 0	11. 8 − 8	12. 19 − 12
13. 17 − 8	14. 11 − 6	15. 8 − 1	16. 15 − 5	17. 20 − 7	18. 12 − 4	19. 5 − 2	20. 16 − 15
21. 5 − 1	22. 7 − 3	23. 16 − 8	24. 17 − 5	25. 8 − 6	26. 8 − 7	27. 13 − 5	28. 17 − 6
29. 5 − 4	30. 20 − 2	31. 20 − 11	32. 19 − 6	33. 18 − 7	34. 13 − 11	35. 20 − 8	36. 9 − 7
37. 14 − 6	38. 20 − 19	39. 18 − 15	40. 15 − 4	41. 12 − 6	42. 10 − 2	43. 17 − 16	44. 17 − 1
45. 16 − 13	46. 19 − 14	47. 15 − 13	48. 18 − 3	49. 20 − 17	50. 17 − 9	51. 11 − 3	52. 19 − 8
53. 14 − 10	54. 19 − 11	55. 0 − 0	56. 19 − 5	57. 13 − 13	58. 3 − 1	59. 20 − 14	60. 7 − 0

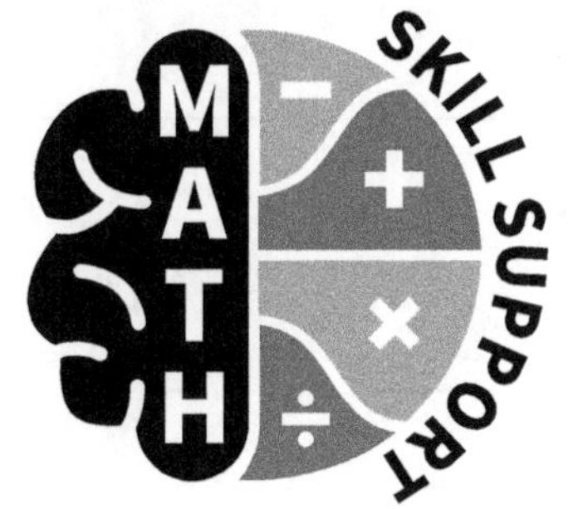

Name: ______________________ Date: ______________

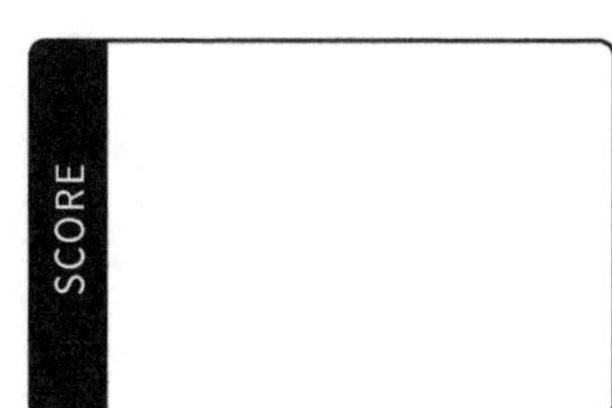

SUBTRACTING WITHIN 20, SET F

				(1) 8 - 3	(2) 11 - 10	(3) 17 - 14	(4) 20 - 18
(5) 19 - 13	(6) 12 - 1	(7) 9 - 5	(8) 13 - 3	(9) 20 - 3	(10) 8 - 5	(11) 12 - 5	(12) 17 - 15
(13) 13 - 2	(14) 12 - 11	(15) 4 - 1	(16) 6 - 2	(17) 11 - 5	(18) 18 - 10	(19) 13 - 4	(20) 14 - 9
(21) 12 - 9	(22) 14 - 1	(23) 3 - 0	(24) 11 - 1	(25) 20 - 16	(26) 15 - 8	(27) 9 - 1	(28) 18 - 8
(29) 14 - 8	(30) 18 - 18	(31) 11 - 2	(32) 13 - 7	(33) 3 - 3	(34) 10 - 10	(35) 17 - 11	(36) 18 - 9
(37) 18 - 11	(38) 20 - 12	(39) 12 - 8	(40) 14 - 0	(41) 1 - 0	(42) 18 - 4	(43) 20 - 9	(44) 4 - 0
(45) 9 - 9	(46) 16 - 11	(47) 16 - 5	(48) 16 - 16	(49) 10 - 4	(50) 18 - 14	(51) 16 - 12	(52) 15 - 11
(53) 18 - 6	(54) 10 - 0	(55) 6 - 6	(56) 17 - 0	(57) 20 - 6	(58) 16 - 2	(59) 12 - 7	(60) 16 - 1

MATH SKILL SUPPORT

Name: ____________________ Date: ____________

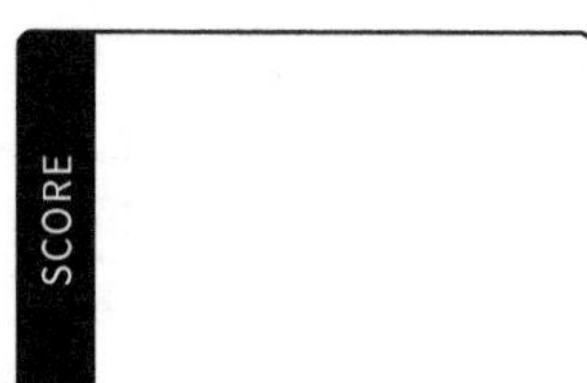
SCORE

1. 2 − 0	2. 14 − 12	3. 15 − 10	4. 18 − 0

5. 17 − 13	6. 20 − 3	7. 15 − 7	8. 18 − 1	9. 9 − 5	10. 6 − 6	11. 13 − 0	12. 19 − 14
13. 15 − 3	14. 3 − 3	15. 15 − 13	16. 17 − 1	17. 14 − 14	18. 16 − 8	19. 14 − 7	20. 15 − 11
21. 20 − 13	22. 18 − 3	23. 19 − 6	24. 10 − 0	25. 20 − 6	26. 9 − 9	27. 18 − 13	28. 17 − 8
29. 16 − 2	30. 18 − 18	31. 15 − 4	32. 10 − 3	33. 5 − 0	34. 14 − 0	35. 11 − 3	36. 13 − 5
37. 4 − 1	38. 4 − 4	39. 8 − 8	40. 8 − 0	41. 9 − 8	42. 7 − 4	43. 16 − 7	44. 9 − 4
45. 16 − 12	46. 3 − 0	47. 11 − 0	48. 6 − 3	49. 19 − 16	50. 0 − 0	51. 14 − 11	52. 9 − 2
53. 10 − 1	54. 7 − 2	55. 10 − 6	56. 16 − 4	57. 19 − 4	58. 18 − 10	59. 8 − 6	60. 19 − 2

SUBTRACTING WITHIN 20, SET H

MATH SKILL SUPPORT

Name: ____________________ Date: ____________

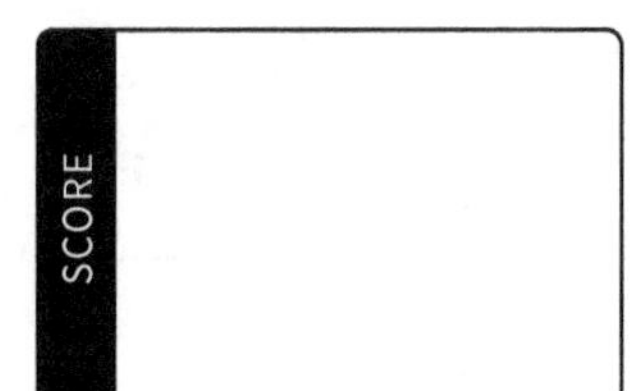

				1. 12 − 4	2. 12 − 12	3. 20 − 2	4. 16 − 14
5. 20 − 11	6. 8 − 5	7. 15 − 2	8. 10 − 10	9. 11 − 5	10. 11 − 9	11. 20 − 20	12. 20 − 17
13. 9 − 6	14. 13 − 3	15. 12 − 0	16. 17 − 9	17. 14 − 4	18. 16 − 10	19. 19 − 17	20. 19 − 13
21. 20 − 8	22. 10 − 8	23. 15 − 6	24. 19 − 18	25. 2 − 2	26. 12 − 5	27. 13 − 13	28. 20 − 12
29. 17 − 11	30. 17 − 16	31. 19 − 8	32. 11 − 4	33. 15 − 5	34. 19 − 12	35. 20 − 10	36. 15 − 14
37. 14 − 5	38. 19 − 19	39. 9 − 3	40. 18 − 15	41. 17 − 2	42. 17 − 10	43. 6 − 2	44. 10 − 4
45. 18 − 2	46. 8 − 7	47. 20 − 0	48. 18 − 7	49. 9 − 0	50. 18 − 8	51. 20 − 7	52. 20 − 19
53. 14 − 3	54. 16 − 5	55. 13 − 10	56. 12 − 7	57. 15 − 9	58. 14 − 13	59. 10 − 2	60. 7 − 1

MATH SKILL SUPPORT

Name: ______________________ Date: ______________

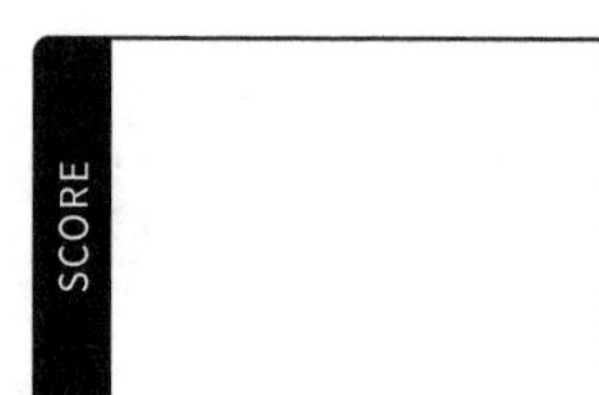
SCORE

				1. $20 - 14$	2. $14 - 2$	3. $11 - 8$	4. $17 - 5$
5. $18 - 4$	6. $19 - 0$	7. $17 - 15$	8. $9 - 1$	9. $18 - 17$	10. $17 - 17$	11. $7 - 3$	12. $7 - 7$
13. $13 - 8$	14. $13 - 9$	15. $18 - 11$	16. $11 - 2$	17. $14 - 8$	18. $12 - 2$	19. $12 - 9$	20. $19 - 7$
21. $16 - 15$	22. $15 - 1$	23. $6 - 5$	24. $12 - 1$	25. $14 - 9$	26. $11 - 1$	27. $14 - 6$	28. $19 - 5$
29. $18 - 16$	30. $2 - 1$	31. $15 - 15$	32. $19 - 3$	33. $6 - 4$	34. $4 - 2$	35. $13 - 11$	36. $19 - 10$
37. $20 - 18$	38. $5 - 3$	39. $14 - 1$	40. $20 - 15$	41. $19 - 1$	42. $17 - 12$	43. $18 - 5$	44. $13 - 6$
45. $11 - 6$	46. $20 - 16$	47. $19 - 11$	48. $6 - 0$	49. $11 - 7$	50. $3 - 1$	51. $10 - 9$	52. $20 - 1$
53. $17 - 3$	54. $9 - 7$	55. $18 - 9$	56. $12 - 8$	57. $14 - 10$	58. $11 - 11$	59. $8 - 3$	60. $20 - 4$

MATH SKILL SUPPORT

Name: ______________________ Date: ______________

SCORE

①	②	③	④
13 -10	20 -19	18 -11	15 - 3

⑤ 16 -15	⑥ 20 -15	⑦ 10 - 1	⑧ 18 -15	⑨ 14 - 5	⑩ 16 - 9	⑪ 19 -18	⑫ 16 -12
⑬ 15 - 7	⑭ 15 -13	⑮ 18 -12	⑯ 20 -13	⑰ 18 - 7	⑱ 18 -18	⑲ 14 -10	⑳ 14 -14
21 20 - 1	22 15 - 9	23 17 -10	24 7 - 4	25 11 - 6	26 6 - 5	27 16 - 0	28 12 -12
29 20 - 9	30 17 - 7	31 20 - 6	32 15 - 5	33 17 -16	34 19 - 3	35 8 - 5	36 7 - 3
37 2 - 1	38 19 -11	39 19 - 6	40 15 - 1	41 16 -14	42 8 - 3	43 7 - 6	44 5 - 3
45 16 - 3	46 14 - 0	47 16 -16	48 20 -16	49 17 -14	50 14 - 2	51 8 - 7	52 20 -10
53 2 - 2	54 12 - 3	55 7 - 7	56 19 -10	57 8 - 4	58 19 -14	59 14 - 1	60 17 - 5

MATH SKILL SUPPORT

Name: ____________________ Date: ____________

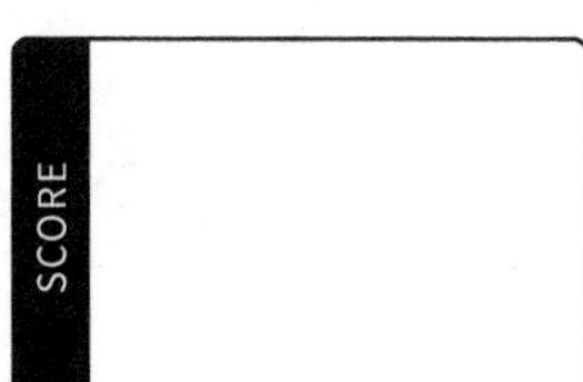
SCORE

				1. 19 − 9	2. 10 − 10	3. 3 − 3	4. 12 − 6
5. 17 − 8	6. 15 − 11	7. 19 − 13	8. 15 − 10	9. 20 − 2	10. 18 − 6	11. 14 − 11	12. 13 − 7
13. 11 − 0	14. 10 − 6	15. 7 − 0	16. 13 − 6	17. 13 − 13	18. 9 − 4	19. 12 − 9	20. 11 − 4
21. 6 − 6	22. 10 − 7	23. 8 − 1	24. 12 − 10	25. 4 − 1	26. 20 − 8	27. 10 − 3	28. 20 − 11
29. 18 − 17	30. 3 − 2	31. 8 − 0	32. 10 − 9	33. 14 − 8	34. 5 − 2	35. 6 − 3	36. 9 − 2
37. 4 − 2	38. 9 − 3	39. 12 − 11	40. 12 − 1	41. 19 − 15	42. 5 − 5	43. 13 − 3	44. 13 − 2
45. 13 − 5	46. 16 − 1	47. 15 − 15	48. 5 − 0	49. 11 − 1	50. 11 − 11	51. 17 − 3	52. 8 − 8
53. 17 − 15	54. 5 − 1	55. 13 − 1	56. 17 − 17	57. 16 − 2	58. 17 − 11	59. 18 − 16	60. 13 − 0

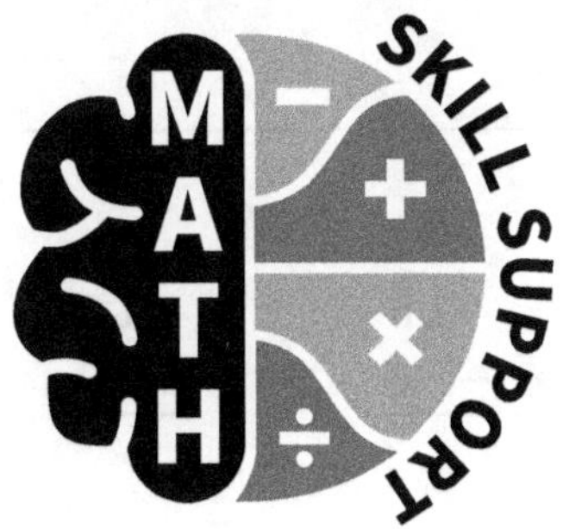

Name: ______________________ Date: ______________

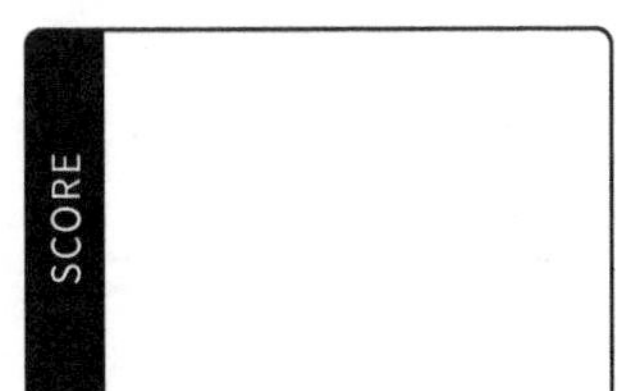

DAY 55

SUBTRACTING WITHIN 20, SET L

				1. 20 - 0	2. 1 - 0	3. 20 - 5	4. 16 - 6
5. 10 - 5	6. 18 - 10	7. 14 - 12	8. 16 - 8	9. 14 - 7	10. 11 - 10	11. 19 - 5	12. 9 - 8
13. 14 - 3	14. 20 - 12	15. 20 - 3	16. 2 - 0	17. 9 - 0	18. 16 - 10	19. 19 - 17	20. 18 - 8
21. 12 - 7	22. 17 - 0	23. 17 - 1	24. 17 - 13	25. 19 - 0	26. 15 - 0	27. 14 - 9	28. 18 - 9
29. 18 - 4	30. 0 - 0	31. 20 - 7	32. 9 - 5	33. 16 - 4	34. 12 - 8	35. 15 - 6	36. 15 - 14
37. 13 - 11	38. 10 - 4	39. 10 - 8	40. 11 - 9	41. 4 - 3	42. 12 - 2	43. 17 - 4	44. 9 - 9
45. 20 - 20	46. 11 - 5	47. 7 - 5	48. 1 - 1	49. 4 - 4	50. 19 - 16	51. 18 - 3	52. 17 - 9
53. 6 - 1	54. 20 - 4	55. 17 - 6	56. 16 - 5	57. 17 - 12	58. 5 - 4	59. 9 - 1	60. 14 - 13

SUBTRACTING WITHIN 20, SET M

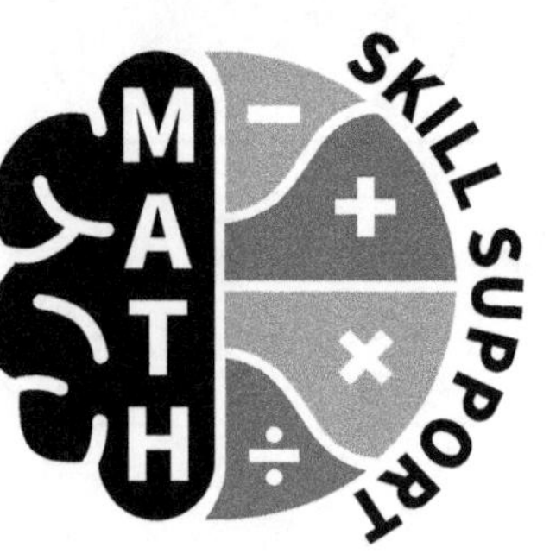

Name: ______________________ Date: ______________

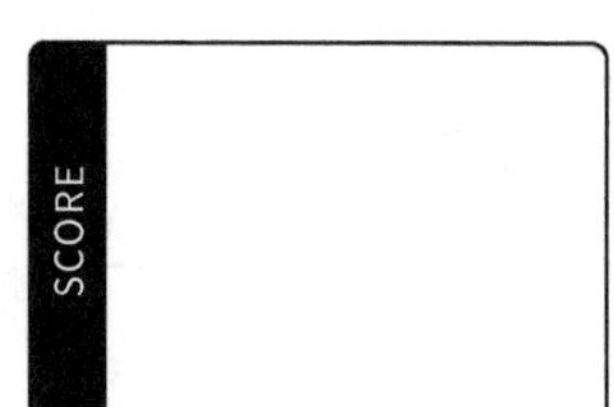

				(1) 14 − 12	(2) 14 − 0	(3) 10 − 2	(4) 12 − 10
(5) 10 − 4	(6) 14 − 11	(7) 20 − 16	(8) 12 − 3	(9) 18 − 18	(10) 12 − 9	(11) 8 − 4	(12) 19 − 17
(13) 13 − 0	(14) 12 − 6	(15) 18 − 9	(16) 14 − 9	(17) 15 − 2	(18) 8 − 0	(19) 3 − 1	(20) 10 − 1
(21) 15 − 0	(22) 8 − 1	(23) 20 − 10	(24) 9 − 5	(25) 12 − 4	(26) 7 − 7	(27) 17 − 7	(28) 18 − 16
(29) 10 − 7	(30) 18 − 1	(31) 14 − 2	(32) 19 − 13	(33) 16 − 10	(34) 13 − 4	(35) 16 − 5	(36) 9 − 1
(37) 17 − 2	(38) 18 − 14	(39) 20 − 6	(40) 14 − 13	(41) 15 − 6	(42) 4 − 2	(43) 10 − 10	(44) 12 − 11
(45) 17 − 16	(46) 19 − 14	(47) 20 − 2	(48) 11 − 8	(49) 10 − 9	(50) 20 − 15	(51) 19 − 5	(52) 14 − 14
(53) 14 − 1	(54) 19 − 10	(55) 13 − 8	(56) 19 − 8	(57) 3 − 3	(58) 13 − 9	(59) 20 − 0	(60) 10 − 3

SUBTRACTING WITHIN 20, SET N

Name: ______________________ Date: ______________

SCORE

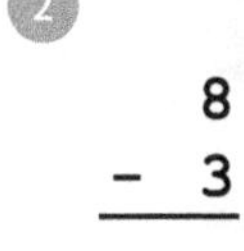
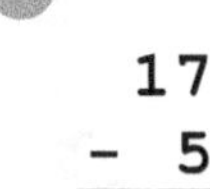
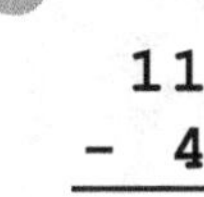

				1. $6 - 4$	2. $8 - 3$	3. $17 - 5$	4. $11 - 4$
5. $13 - 2$	6. $16 - 6$	7. $13 - 6$	8. $15 - 10$	9. $5 - 1$	10. $8 - 7$	11. $11 - 7$	12. $16 - 3$
13. $4 - 3$	14. $15 - 15$	15. $10 - 8$	16. $13 - 13$	17. $15 - 11$	18. $11 - 3$	19. $18 - 7$	20. $20 - 1$
21. $4 - 0$	22. $15 - 1$	23. $17 - 12$	24. $19 - 2$	25. $18 - 13$	26. $9 - 6$	27. $15 - 7$	28. $8 - 2$
29. $16 - 11$	30. $20 - 19$	31. $9 - 3$	32. $20 - 20$	33. $20 - 7$	34. $16 - 4$	35. $17 - 11$	36. $16 - 7$
37. $5 - 4$	38. $7 - 0$	39. $16 - 8$	40. $7 - 1$	41. $16 - 9$	42. $20 - 17$	43. $18 - 12$	44. $15 - 8$
45. $16 - 16$	46. $14 - 8$	47. $18 - 8$	48. $12 - 7$	49. $10 - 0$	50. $17 - 8$	51. $9 - 7$	52. $6 - 2$
53. $5 - 0$	54. $13 - 10$	55. $19 - 3$	56. $9 - 0$	57. $19 - 6$	58. $13 - 1$	59. $18 - 6$	60. $16 - 1$

MATH SKILL SUPPORT

Name: ____________________ Date: ____________

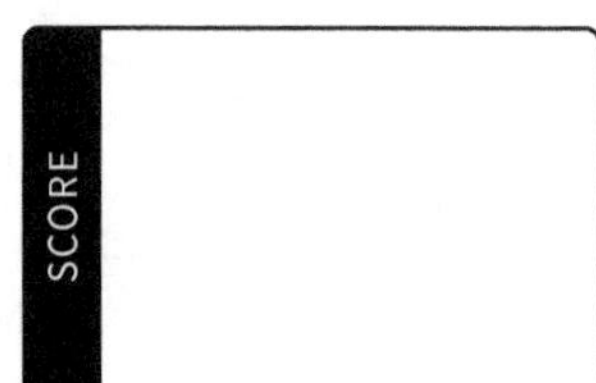
SCORE

				1) 8 - 5	2) 6 - 0	3) 5 - 2	4) 20 - 3
5) 15 - 9	6) 20 - 4	7) 12 - 0	8) 20 - 13	9) 18 - 0	10) 12 - 5	11) 14 - 5	12) 18 - 11
13) 10 - 6	14) 18 - 3	15) 9 - 9	16) 16 - 13	17) 12 - 12	18) 11 - 11	19) 8 - 8	20) 1 - 1
21) 19 - 19	22) 18 - 2	23) 10 - 5	24) 19 - 16	25) 2 - 2	26) 7 - 3	27) 15 - 5	28) 15 - 3
29) 17 - 10	30) 5 - 3	31) 12 - 2	32) 11 - 0	33) 20 - 12	34) 4 - 1	35) 11 - 1	36) 1 - 0
37) 20 - 11	38) 19 - 18	39) 16 - 2	40) 18 - 4	41) 13 - 11	42) 14 - 4	43) 15 - 14	44) 4 - 4
45) 19 - 7	46) 19 - 4	47) 13 - 12	48) 13 - 3	49) 7 - 5	50) 20 - 18	51) 18 - 17	52) 17 - 14
53) 20 - 5	54) 18 - 5	55) 0 - 0	56) 19 - 12	57) 19 - 15	58) 14 - 3	59) 15 - 4	60) 17 - 17

SKILL SUPPORT MATH

Name: ____________________ Date: ____________________

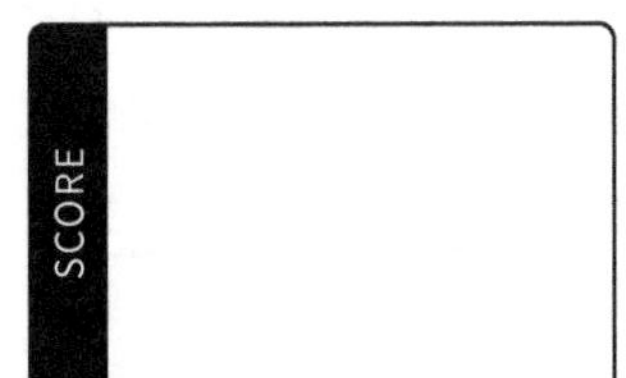

#	Problem	#	Problem	#	Problem	#	Problem
1	20 - 18	2	17 - 14	3	14 - 5	4	18 - 3
5	8 - 3	6	11 - 8	7	1 - 1	8	16 - 8
9	5 - 5	10	6 - 1	11	12 - 0	12	13 - 8
13	14 - 3	14	13 - 11	15	15 - 13	16	10 - 5
17	8 - 4	18	18 - 2	19	19 - 8	20	7 - 4
21	16 - 3	22	16 - 0	23	11 - 1	24	14 - 12
25	17 - 6	26	20 - 2	27	14 - 4	28	17 - 8
29	20 - 19	30	9 - 7	31	9 - 9	32	12 - 5
33	11 - 5	34	15 - 1	35	11 - 6	36	12 - 2
37	15 - 5	38	17 - 11	39	8 - 1	40	10 - 10
41	18 - 18	42	15 - 14	43	15 - 6	44	11 - 9
45	18 - 16	46	20 - 16	47	18 - 15	48	11 - 10
49	16 - 7	50	6 - 5	51	7 - 0	52	19 - 6
53	19 - 12	54	18 - 13	55	12 - 7	56	12 - 9
57	15 - 4	58	12 - 12	59	15 - 3	60	19 - 3

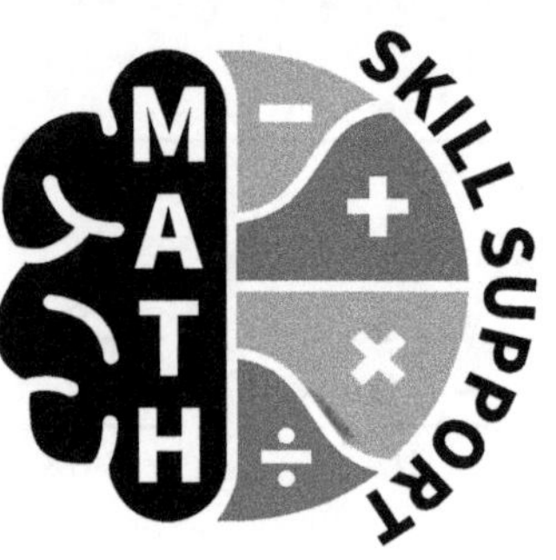

Name: ____________________ Date: ____________

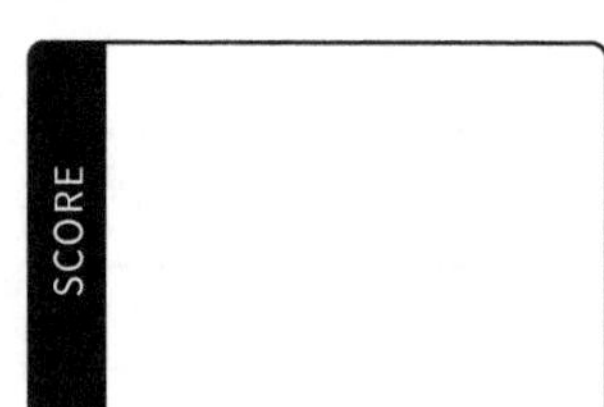

				① 2 − 1	② 18 − 12	③ 14 − 6	④ 20 − 17
⑤ 6 − 0	⑥ 10 − 8	⑦ 11 − 0	⑧ 14 − 2	⑨ 11 − 3	⑩ 16 − 2	⑪ 16 − 13	⑫ 13 − 5
⑬ 4 − 4	⑭ 3 − 1	⑮ 10 − 6	⑯ 18 − 10	⑰ 11 − 2	⑱ 13 − 12	⑲ 17 − 16	⑳ 18 − 14
21. 17 − 7	22. 9 − 1	23. 8 − 5	24. 20 − 9	25. 12 − 11	26. 19 − 7	27. 17 − 17	28. 17 − 1
29. 13 − 0	30. 14 − 0	31. 13 − 13	32. 4 − 0	33. 19 − 19	34. 4 − 2	35. 18 − 4	36. 20 − 7
37. 9 − 6	38. 16 − 15	39. 17 − 13	40. 20 − 3	41. 15 − 9	42. 16 − 1	43. 16 − 14	44. 18 − 9
45. 19 − 17	46. 19 − 2	47. 19 − 5	48. 19 − 15	49. 16 − 10	50. 2 − 0	51. 11 − 7	52. 9 − 8
53. 12 − 1	54. 5 − 3	55. 7 − 1	56. 11 − 11	57. 16 − 12	58. 6 − 2	59. 14 − 8	60. 15 − 2

SUBTRACTING WITHIN 20, SET R

MATH SKILL SUPPORT

Name: ____________________ Date: ____________________

				1. 4 − 3	2. 10 − 3	3. 18 − 7	4. 20 − 0
5. 7 − 2	6. 14 − 14	7. 20 − 10	8. 7 − 3	9. 19 − 18	10. 17 − 3	11. 18 − 5	12. 20 − 20
13. 20 − 6	14. 11 − 4	15. 20 − 8	16. 20 − 15	17. 13 − 7	18. 12 − 4	19. 16 − 9	20. 19 − 0
21. 12 − 10	22. 17 − 4	23. 15 − 12	24. 19 − 1	25. 10 − 4	26. 20 − 1	27. 19 − 11	28. 14 − 11
29. 3 − 2	30. 10 − 2	31. 16 − 11	32. 3 − 0	33. 9 − 3	34. 12 − 8	35. 9 − 0	36. 19 − 9
37. 18 − 1	38. 10 − 9	39. 7 − 5	40. 14 − 10	41. 20 − 11	42. 1 − 0	43. 20 − 5	44. 12 − 6
45. 10 − 7	46. 12 − 3	47. 8 − 6	48. 7 − 7	49. 15 − 8	50. 14 − 7	51. 13 − 9	52. 17 − 0
53. 5 − 4	54. 8 − 0	55. 18 − 8	56. 16 − 5	57. 6 − 4	58. 4 − 1	59. 19 − 10	60. 15 − 15

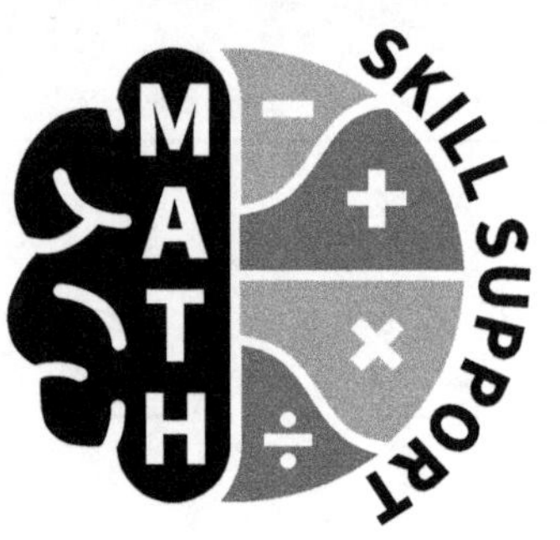

Name: ____________________ Date: ____________________

				(1) 7 − 5	(2) 15 − 5	(3) 13 − 0	(4) 11 − 10
(5) 1 − 1	(6) 11 − 4	(7) 4 − 3	(8) 14 − 4	(9) 8 − 0	(10) 9 − 6	(11) 8 − 6	(12) 20 − 11
(13) 8 − 5	(14) 10 − 5	(15) 8 − 3	(16) 16 − 1	(17) 18 − 2	(18) 18 − 9	(19) 20 − 2	(20) 14 − 10
(21) 9 − 5	(22) 16 − 6	(23) 19 − 2	(24) 11 − 6	(25) 20 − 6	(26) 12 − 6	(27) 17 − 16	(28) 10 − 7
(29) 9 − 0	(30) 6 − 3	(31) 9 − 8	(32) 13 − 6	(33) 19 − 0	(34) 16 − 5	(35) 19 − 11	(36) 19 − 17
(37) 12 − 9	(38) 5 − 0	(39) 15 − 14	(40) 16 − 2	(41) 7 − 2	(42) 10 − 3	(43) 3 − 2	(44) 13 − 4
(45) 8 − 7	(46) 16 − 13	(47) 13 − 3	(48) 14 − 2	(49) 8 − 2	(50) 20 − 9	(51) 10 − 2	(52) 19 − 1
(53) 14 − 14	(54) 20 − 7	(55) 17 − 8	(56) 10 − 8	(57) 20 − 18	(58) 19 − 4	(59) 13 − 5	(60) 11 − 7

SUBTRACTING WITHIN 20, SET T

Name: ____________________ Date: ____________

1. 17 − 0
2. 14 − 9
3. 9 − 4
4. 18 − 10
5. 13 − 13
6. 17 − 12
7. 7 − 0
8. 19 − 8
9. 20 − 10
10. 17 − 17
11. 12 − 7
12. 16 − 9
13. 16 − 12
14. 18 − 0
15. 12 − 11
16. 20 − 0
17. 15 − 1
18. 4 − 2
19. 8 − 4
20. 14 − 7
21. 13 − 11
22. 9 − 2
23. 17 − 11
24. 15 − 7
25. 17 − 13
26. 5 − 4
27. 18 − 6
28. 20 − 16
29. 20 − 13
30. 15 − 13
31. 5 − 2
32. 18 − 3
33. 14 − 5
34. 12 − 2
35. 19 − 10
36. 10 − 1
37. 16 − 11
38. 17 − 7
39. 2 − 2
40. 3 − 1
41. 11 − 3
42. 15 − 6
43. 17 − 5
44. 16 − 15
45. 18 − 5
46. 18 − 11
47. 18 − 15
48. 7 − 1
49. 16 − 14
50. 10 − 4
51. 16 − 16
52. 15 − 11
53. 18 − 14
54. 15 − 4
55. 9 − 9
56. 19 − 5
57. 19 − 13
58. 4 − 4
59. 14 − 12
60. 18 − 16

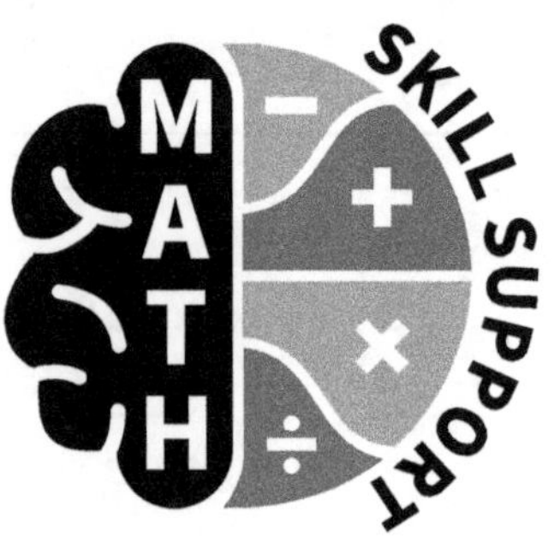

Name: ____________________ Date: ____________

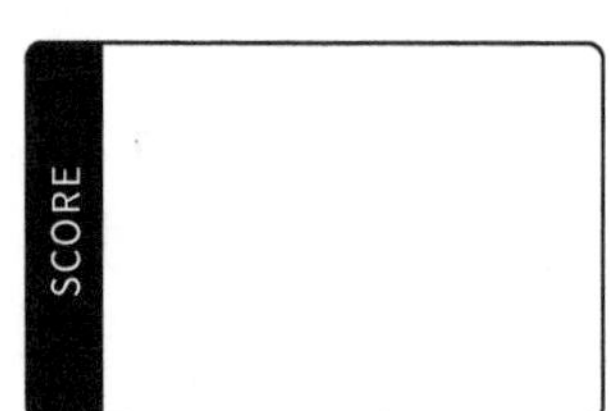

				(1) 7 − 3	(2) 18 − 7	(3) 17 − 9	(4) 16 − 8
(5) 13 − 10	(6) 19 − 15	(7) 12 − 8	(8) 5 − 1	(9) 3 − 3	(10) 15 − 9	(11) 4 − 1	(12) 20 − 1
(13) 15 − 3	(14) 9 − 1	(15) 16 − 4	(16) 15 − 15	(17) 12 − 5	(18) 7 − 6	(19) 2 − 0	(20) 12 − 3
(21) 20 − 5	(22) 20 − 15	(23) 19 − 7	(24) 14 − 6	(25) 12 − 10	(26) 13 − 9	(27) 12 − 0	(28) 20 − 14
(29) 19 − 18	(30) 17 − 4	(31) 8 − 8	(32) 18 − 17	(33) 11 − 11	(34) 6 − 1	(35) 17 − 6	(36) 16 − 0
(37) 5 − 3	(38) 7 − 7	(39) 15 − 0	(40) 6 − 5	(41) 5 − 5	(42) 12 − 1	(43) 9 − 7	(44) 17 − 2
(45) 18 − 4	(46) 15 − 10	(47) 19 − 12	(48) 14 − 11	(49) 18 − 18	(50) 20 − 19	(51) 17 − 15	(52) 11 − 9
(53) 17 − 14	(54) 8 − 1	(55) 10 − 6	(56) 16 − 10	(57) 16 − 7	(58) 11 − 0	(59) 20 − 20	(60) 20 − 17

MATH SKILL SUPPORT

Name: ____________________ Date: ____________________

SCORE

DAY 65

MIXED REVIEW: EVENS ONLY, SET A

				1) 2 + 14	2) 0 + 2	3) 6 − 4	4) 20 − 14
5) 4 − 2	6) 6 + 4	7) 20 − 8	8) 6 + 10	9) 2 + 4	10) 4 + 12	11) 20 − 6	12) 10 − 2
13) 6 + 14	14) 8 + 12	15) 10 − 10	16) 4 + 0	17) 18 − 16	18) 18 − 0	19) 14 + 2	20) 16 − 4
21) 16 − 6	22) 18 − 10	23) 20 − 12	24) 12 + 2	25) 6 − 0	26) 18 − 14	27) 12 − 8	28) 8 + 8
29) 12 + 8	30) 14 + 4	31) 12 + 0	32) 14 − 2	33) 2 − 0	34) 20 − 16	35) 20 − 0	36) 4 + 16
37) 0 + 16	38) 6 + 6	39) 8 + 0	40) 12 − 10	41) 0 + 10	42) 14 − 6	43) 18 + 2	44) 16 − 10
45) 18 − 18	46) 14 − 4	47) 6 + 2	48) 2 + 18	49) 8 + 2	50) 12 + 4	51) 14 − 0	52) 16 + 4
53) 0 + 20	54) 12 − 0	55) 0 + 0	56) 8 − 4	57) 18 − 4	58) 2 + 16	59) 18 − 12	60) 2 + 6

MIXED REVIEW: EVENS ONLY, SET B

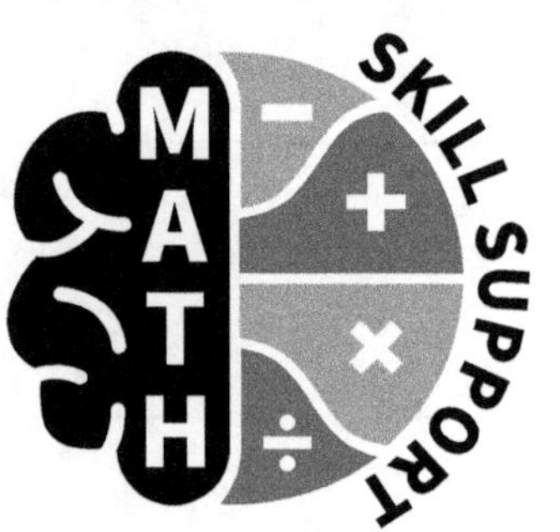

Name: ____________________ Date: ____________________

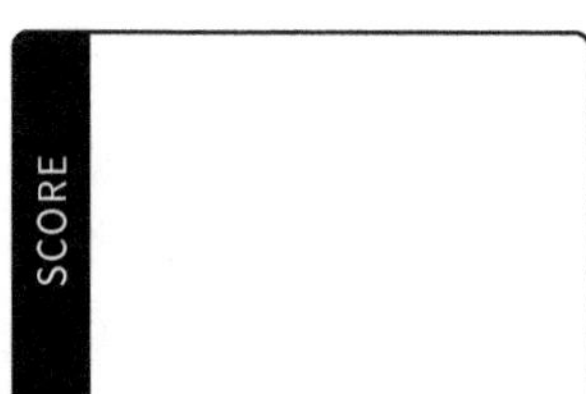

				(1) 0 +18	(2) 18 −14	(3) 16 − 0	(4) 10 + 2
(5) 2 + 4	(6) 14 − 8	(7) 18 −18	(8) 2 +16	(9) 2 + 0	(10) 12 + 6	(11) 12 −10	(12) 2 +12
(13) 4 +16	(14) 2 − 0	(15) 16 −16	(16) 18 − 6	(17) 16 + 4	(18) 8 − 8	(19) 18 −10	(20) 16 + 0
(21) 6 + 0	(22) 18 − 4	(23) 0 +20	(24) 0 +14	(25) 14 − 0	(26) 0 +12	(27) 4 +10	(28) 12 + 2
(29) 20 − 2	(30) 20 −10	(31) 20 − 4	(32) 20 −14	(33) 6 − 2	(34) 4 + 4	(35) 4 +14	(36) 6 + 2
(37) 16 − 4	(38) 8 +12	(39) 6 + 8	(40) 12 − 8	(41) 16 − 6	(42) 16 −12	(43) 10 −10	(44) 0 +16
(45) 14 + 6	(46) 14 − 2	(47) 10 − 0	(48) 12 + 0	(49) 8 + 2	(50) 10 − 4	(51) 14 −10	(52) 2 + 8
(53) 10 + 4	(54) 6 − 0	(55) 10 − 8	(56) 4 + 8	(57) 6 + 4	(58) 8 − 6	(59) 10 + 0	(60) 18 −12

Name: ____________________ Date: ____________

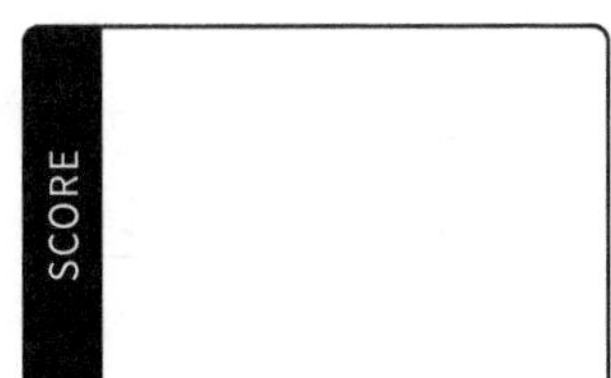

				1 10 -10	2 10 + 8	3 2 +14	4 14 - 4
5 8 - 0	6 18 - 8	7 4 + 2	8 0 + 2	9 0 +18	10 4 - 4	11 2 +18	12 8 - 8
13 20 -14	14 16 - 2	15 18 + 0	16 2 +10	17 6 + 2	18 18 -18	19 6 - 6	20 4 + 4
21 10 + 2	22 0 +10	23 12 - 6	24 8 + 2	25 6 - 0	26 14 - 6	27 6 +12	28 0 +16
29 12 - 4	30 12 + 6	31 2 - 2	32 12 -12	33 16 - 6	34 8 + 0	35 12 - 8	36 6 +10
37 10 + 0	38 18 -14	39 4 - 2	40 2 + 0	41 8 + 6	42 20 -16	43 4 + 6	44 14 - 2
45 6 + 6	46 0 + 8	47 6 +14	48 20 - 6	49 12 + 4	50 20 - 2	51 10 - 4	52 0 +14
53 18 - 0	54 6 - 4	55 20 + 0	56 8 - 4	57 10 - 2	58 4 +10	59 2 + 8	60 12 - 0

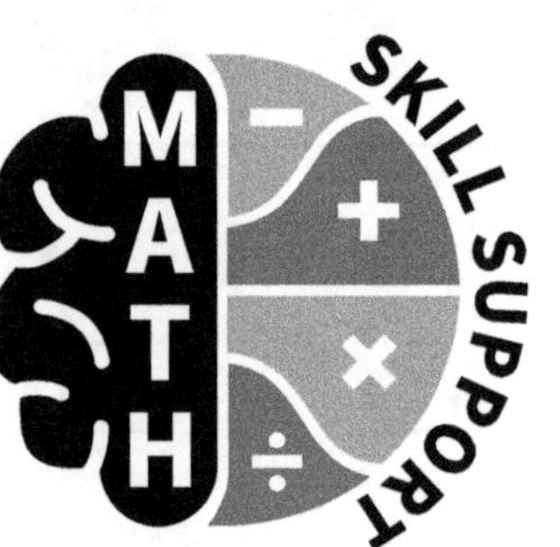

Name: ____________________ Date: ____________

1. 16 − 0	2. 2 + 10	3. 4 + 2	4. 2 + 12

5. 20 − 12	6. 18 − 6	7. 20 − 4	8. 4 − 2	9. 10 − 10	10. 4 + 0	11. 14 − 6	12. 12 − 8
13. 8 − 8	14. 16 − 16	15. 2 + 6	16. 8 − 0	17. 8 + 10	18. 20 − 8	19. 18 − 4	20. 12 + 8
21. 6 + 6	22. 12 − 4	23. 10 − 0	24. 18 + 0	25. 0 + 0	26. 2 + 16	27. 20 − 18	28. 0 + 14
29. 10 + 4	30. 0 + 4	31. 20 − 0	32. 10 + 0	33. 8 − 2	34. 18 − 8	35. 0 + 20	36. 14 − 8
37. 18 − 18	38. 4 + 4	39. 8 + 2	40. 12 − 10	41. 16 + 2	42. 6 − 4	43. 2 + 2	44. 6 + 4
45. 10 − 2	46. 18 + 2	47. 12 − 6	48. 6 − 0	49. 0 + 6	50. 14 − 2	51. 8 + 8	52. 4 + 14
53. 18 − 2	54. 8 − 6	55. 12 + 2	56. 2 + 14	57. 16 + 4	58. 14 + 4	59. 14 − 0	60. 12 + 6

MATH SKILL SUPPORT

Name: ____________________ Date: ____________________

MIXED REVIEW: EVENS ONLY, SET E

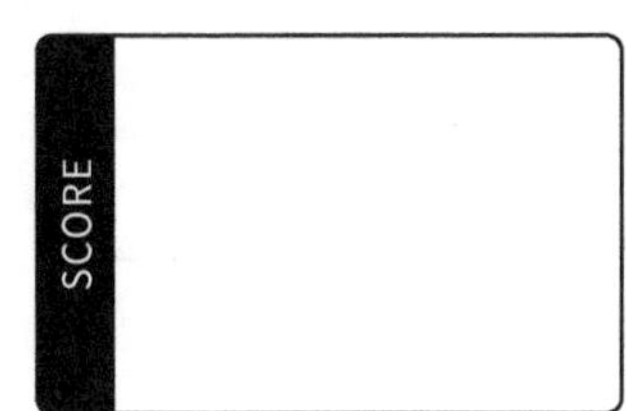

				1 14 -10	2 8 - 6	3 2 - 0	4 10 - 4
5 2 + 2	6 16 - 4	7 0 + 4	8 10 + 2	9 16 + 2	10 12 + 4	11 16 + 0	12 12 + 0
13 12 + 2	14 20 -20	15 6 + 0	16 2 + 4	17 6 - 4	18 16 + 4	19 14 -12	20 16 -16
21 8 - 0	22 2 +14	23 4 +16	24 18 - 6	25 0 +12	26 18 - 8	27 14 - 0	28 20 -18
29 20 - 6	30 12 - 2	31 2 - 2	32 12 - 8	33 10 + 6	34 0 +14	35 10 - 0	36 12 + 8
37 4 + 2	38 6 + 2	39 4 +10	40 12 - 4	41 18 + 2	42 16 -14	43 12 + 6	44 0 + 2
45 12 -10	46 6 - 0	47 0 +10	48 2 + 0	49 6 +12	50 16 - 0	51 6 + 8	52 10 - 2
53 6 +14	54 6 - 6	55 4 - 2	56 0 + 0	57 0 - 0	58 4 +12	59 20 - 0	60 18 -12

DAY 70

MIXED REVIEW: ODDS ONLY, SET A

MATH SKILL SUPPORT

Name: ____________________ Date: ____________

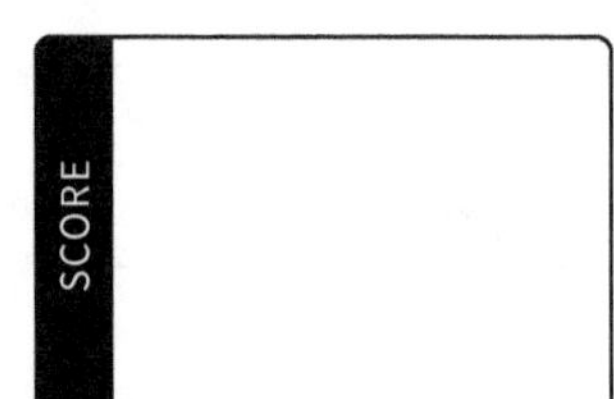

				1. 7 + 13	2. 1 + 7	3. 9 + 3	4. 17 - 7
5. 13 - 5	6. 15 - 1	7. 11 + 5	8. 19 - 13	9. 9 - 9	10. 3 - 1	11. 3 - 3	12. 15 - 9
13. 7 + 5	14. 7 - 3	15. 17 - 15	16. 3 + 7	17. 5 - 1	18. 19 - 19	19. 15 - 13	20. 1 + 3
21. 13 + 5	22. 15 + 3	23. 17 - 1	24. 9 + 5	25. 11 + 1	26. 3 + 1	27. 7 + 3	28. 3 + 3
29. 9 + 7	30. 1 + 15	31. 17 - 11	32. 17 + 1	33. 11 - 1	34. 17 - 3	35. 5 + 1	36. 7 + 9
37. 19 - 9	38. 11 - 3	39. 15 - 7	40. 1 + 13	41. 11 - 11	42. 5 + 5	43. 1 + 1	44. 9 - 1
45. 11 - 7	46. 3 + 11	47. 9 - 5	48. 13 - 9	49. 17 - 5	50. 11 + 3	51. 13 - 7	52. 13 - 1
53. 9 + 11	54. 1 - 1	55. 7 + 7	56. 9 + 1	57. 19 - 7	58. 15 + 5	59. 1 + 19	60. 3 + 17

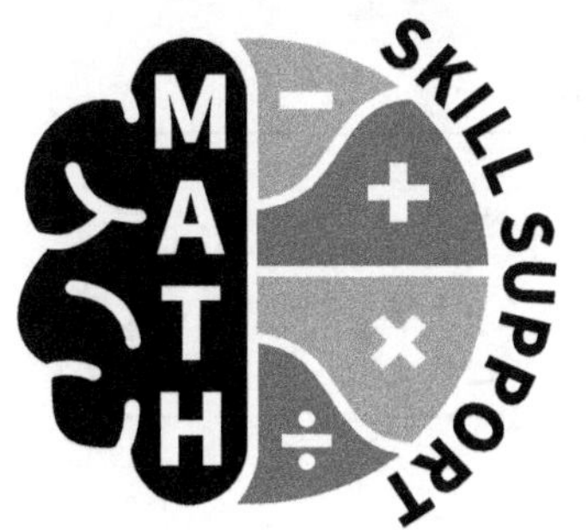

Name: ____________________ Date: ____________________

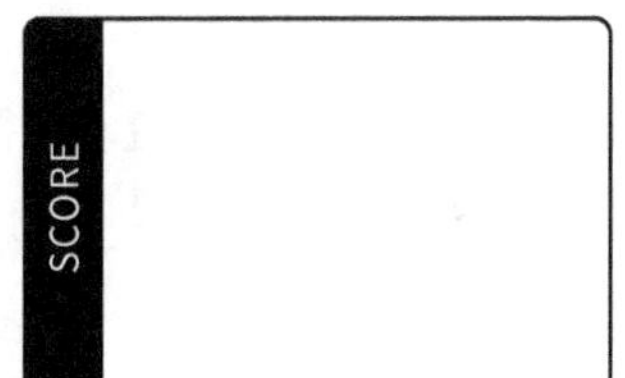

DAY 71

MIXED REVIEW: ODDS ONLY, SET B

1. $1 + 3$
2. $17 - 3$
3. $17 - 5$
4. $11 - 5$
5. $3 + 7$
6. $7 - 1$
7. $3 + 13$
8. $1 + 1$
9. $13 + 1$
10. $17 - 1$
11. $3 - 3$
12. $13 + 7$
13. $13 - 11$
14. $17 + 3$
15. $11 + 9$
16. $7 + 3$
17. $15 - 5$
18. $9 - 7$
19. $1 + 11$
20. $11 + 1$
21. $5 + 9$
22. $9 - 3$
23. $19 - 1$
24. $15 - 11$
25. $17 + 1$
26. $11 + 5$
27. $5 + 1$
28. $7 + 1$
29. $3 + 1$
30. $13 - 13$
31. $1 + 13$
32. $17 - 17$
33. $9 + 5$
34. $19 - 17$
35. $5 + 3$
36. $3 + 15$
37. $17 - 15$
38. $13 - 3$
39. $11 - 11$
40. $19 - 9$
41. $9 + 1$
42. $5 + 5$
43. $5 - 1$
44. $5 + 7$
45. $15 - 13$
46. $19 - 3$
47. $9 - 9$
48. $1 + 5$
49. $11 - 3$
50. $7 - 3$
51. $19 - 15$
52. $3 + 5$
53. $1 + 7$
54. $13 - 9$
55. $13 - 7$
56. $7 - 7$
57. $15 + 3$
58. $15 + 5$
59. $19 - 5$
60. $7 + 9$

DAY 72

MIXED REVIEW: ODDS ONLY, SET C

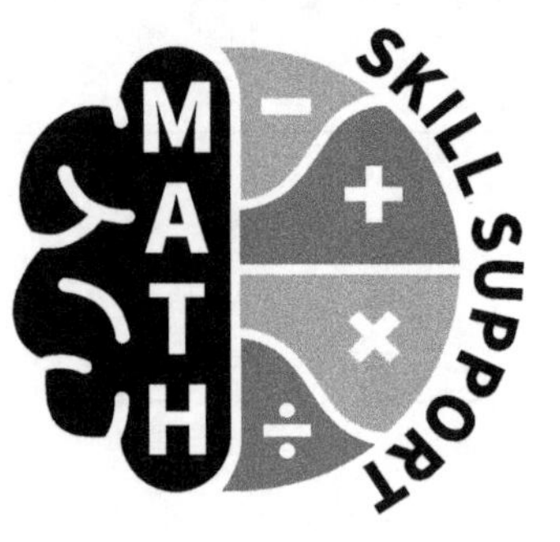

Name: ____________________ Date: ____________

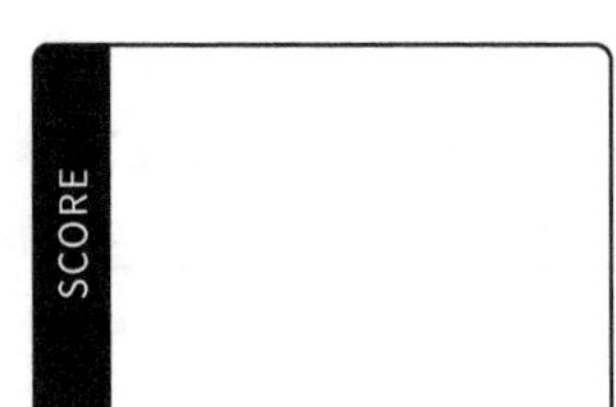

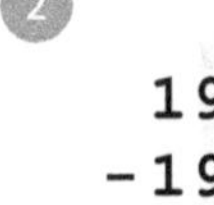

				(1) 1 + 3	(2) 19 − 19	(3) 17 − 3	(4) 13 − 5
(5) 5 + 1	(6) 19 − 17	(7) 13 − 9	(8) 7 + 13	(9) 5 + 3	(10) 9 + 9	(11) 17 − 9	(12) 15 − 11
(13) 7 + 5	(14) 1 + 13	(15) 7 + 11	(16) 17 − 13	(17) 9 − 3	(18) 15 − 3	(19) 17 + 1	(20) 15 + 5
(21) 7 + 9	(22) 11 − 5	(23) 1 + 9	(24) 1 + 17	(25) 15 − 13	(26) 19 − 15	(27) 5 + 15	(28) 15 − 5
(29) 11 − 11	(30) 15 − 7	(31) 7 + 7	(32) 7 + 3	(33) 9 + 3	(34) 3 + 3	(35) 9 + 7	(36) 7 − 5
(37) 19 − 3	(38) 7 − 3	(39) 15 + 1	(40) 5 − 5	(41) 7 + 1	(42) 1 + 15	(43) 15 − 9	(44) 9 + 11
(45) 5 − 3	(46) 11 − 1	(47) 7 − 1	(48) 19 − 5	(49) 15 + 3	(50) 11 − 3	(51) 9 − 9	(52) 17 − 7
(53) 5 + 9	(54) 1 + 7	(55) 3 + 5	(56) 15 − 1	(57) 19 − 1	(58) 1 + 5	(59) 1 + 19	(60) 9 + 5

MIXED REVIEW: ODDS ONLY, SET D

MATH SKILL SUPPORT

Name: ____________________ Date: ____________

SCORE

				1. $5 + 7$	2. $17 - 7$	3. $1 + 13$	4. $17 - 1$
5. $9 - 7$	6. $5 - 3$	7. $3 + 11$	8. $15 - 5$	9. $9 + 3$	10. $1 + 5$	11. $1 + 19$	12. $9 + 11$
13. $1 + 15$	14. $5 + 5$	15. $7 - 3$	16. $19 - 7$	17. $13 - 11$	18. $15 - 15$	19. $9 + 9$	20. $1 + 1$
21. $17 - 13$	22. $15 - 1$	23. $3 + 17$	24. $1 + 9$	25. $11 + 1$	26. $5 + 1$	27. $5 + 13$	28. $11 - 3$
29. $3 + 9$	30. $7 + 3$	31. $7 + 13$	32. $9 - 9$	33. $19 - 1$	34. $5 + 11$	35. $3 - 1$	36. $15 + 1$
37. $13 + 7$	38. $17 - 11$	39. $11 + 5$	40. $11 - 5$	41. $7 - 7$	42. $13 - 1$	43. $17 - 9$	44. $15 - 11$
45. $11 - 7$	46. $11 + 9$	47. $17 - 15$	48. $15 - 13$	49. $3 + 13$	50. $7 + 9$	51. $1 - 1$	52. $13 + 3$
53. $1 + 7$	54. $7 - 1$	55. $19 - 5$	56. $9 + 5$	57. $3 + 5$	58. $9 - 5$	59. $19 - 13$	60. $13 - 7$

MATH SKILL SUPPORT

Name: ______________________ Date: ______________

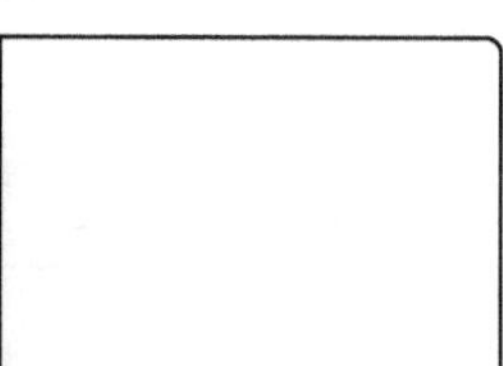
SCORE

#	Problem	#	Problem	#	Problem	#	Problem
1	7 − 3	2	7 + 13	3	11 + 3	4	5 + 9
5	19 − 15	6	13 − 5	7	11 + 5	8	15 − 7
9	13 + 5	10	3 + 15	11	9 − 5	12	15 − 13
13	1 + 17	14	17 + 3	15	3 + 7	16	1 + 3
17	3 + 11	18	17 − 5	19	17 + 1	20	9 + 7
21	15 − 1	22	11 − 3	23	17 − 17	24	11 − 1
25	1 + 9	26	7 + 11	27	3 + 13	28	17 − 1
29	11 + 7	30	9 + 9	31	7 + 7	32	7 + 5
33	13 − 7	34	7 + 3	35	13 − 3	36	7 − 1
37	17 − 13	38	13 − 13	39	15 − 5	40	15 + 5
41	17 − 11	42	17 − 3	43	11 + 1	44	11 − 5
45	7 + 1	46	9 + 11	47	1 − 1	48	11 − 7
49	1 + 1	50	9 − 9	51	13 + 7	52	19 − 13
53	17 − 9	54	3 + 1	55	7 − 7	56	3 − 3
57	19 − 9	58	5 + 3	59	17 − 7	60	3 + 17

MATH SKILL SUPPORT

Name: ____________________ Date: ____________________

SCORE

MIXED REVIEW: ADDING AND SUBTRACTING 0-20, SET A

1. 5 − 1	2. 7 + 10	3. 2 + 3	4. 8 + 2

5. 7 + 9	6. 3 + 13	7. 19 − 4	8. 18 + 0	9. 14 + 2	10. 8 + 5	11. 18 − 3	12. 10 − 7
13. 3 − 2	14. 13 − 4	15. 4 + 16	16. 0 + 4	17. 17 − 12	18. 18 − 13	19. 5 + 15	20. 10 + 1
21. 5 − 5	22. 19 − 8	23. 1 + 6	24. 18 + 1	25. 6 + 11	26. 0 + 11	27. 17 − 2	28. 6 + 14
29. 3 − 0	30. 5 + 12	31. 13 − 8	32. 5 − 2	33. 10 − 6	34. 7 + 2	35. 17 + 2	36. 8 − 6
37. 15 + 0	38. 0 + 2	39. 17 − 11	40. 1 + 4	41. 6 + 1	42. 0 + 20	43. 9 − 1	44. 0 + 19
45. 1 + 5	46. 17 − 8	47. 15 − 3	48. 18 − 4	49. 3 − 3	50. 17 + 0	51. 16 + 1	52. 14 − 11
53. 18 − 15	54. 19 − 3	55. 19 − 12	56. 16 − 3	57. 11 − 6	58. 16 − 13	59. 13 + 1	60. 19 − 7

DAY 76

MIXED REVIEW: ADDING AND SUBTRACTING 0-20, SET B

MATH SKILL SUPPORT

Name: ____________________ Date: ____________

SCORE

				(1) 17 - 4	(2) 14 - 7	(3) 15 - 8	(4) 1 + 3
(5) 13 - 2	(6) 14 - 3	(7) 8 +12	(8) 11 + 3	(9) 17 -15	(10) 19 -13	(11) 13 -13	(12) 6 + 6
(13) 3 +11	(14) 17 -16	(15) 4 + 2	(16) 14 + 5	(17) 12 - 6	(18) 2 +11	(19) 12 -10	(20) 6 - 0
(21) 13 + 5	(22) 10 +10	(23) 0 +18	(24) 15 + 1	(25) 11 - 8	(26) 5 + 9	(27) 18 + 2	(28) 0 +12
(29) 13 + 3	(30) 9 - 6	(31) 2 - 2	(32) 16 -14	(33) 16 + 2	(34) 10 - 1	(35) 14 - 4	(36) 4 + 6
(37) 20 -19	(38) 12 - 8	(39) 10 - 4	(40) 4 + 4	(41) 11 + 6	(42) 2 +10	(43) 12 - 1	(44) 2 +14
(45) 12 -11	(46) 17 -10	(47) 8 + 3	(48) 0 + 1	(49) 11 -10	(50) 7 - 0	(51) 1 +18	(52) 16 + 3
(53) 19 -18	(54) 14 -12	(55) 2 +17	(56) 2 + 2	(57) 5 + 2	(58) 15 + 2	(59) 7 - 7	(60) 19 -15

MATH SKILL SUPPORT

Name: ____________________ Date: ____________________

MIXED REVIEW: ADDING AND SUBTRACTING 0-20, SET C

SCORE

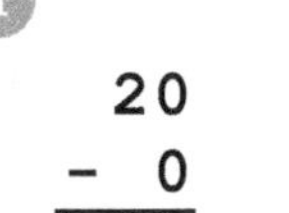

1. $20 - 0$	2. $16 - 0$	3. $18 - 17$	4. $4 + 12$

5. $2 + 8$	6. $9 + 5$	7. $20 - 1$	8. $3 + 1$	9. $5 + 3$	10. $12 + 7$	11. $4 - 0$	12. $13 - 7$
13. $7 - 1$	14. $7 + 3$	15. $14 + 3$	16. $11 - 5$	17. $5 - 0$	18. $10 + 6$	19. $13 + 2$	20. $12 - 3$
21. $15 + 3$	22. $11 + 8$	23. $20 - 9$	24. $0 + 10$	25. $11 + 9$	26. $9 + 6$	27. $15 - 14$	28. $1 + 14$
29. $6 + 10$	30. $8 + 11$	31. $8 - 0$	32. $5 + 6$	33. $2 - 1$	34. $11 - 1$	35. $5 + 11$	36. $18 - 2$
37. $5 + 7$	38. $4 - 1$	39. $12 - 4$	40. $19 - 6$	41. $9 + 11$	42. $5 - 4$	43. $18 - 16$	44. $20 - 12$
45. $17 - 17$	46. $4 + 0$	47. $13 - 1$	48. $4 + 5$	49. $7 - 3$	50. $1 + 11$	51. $13 - 6$	52. $11 + 7$
53. $2 + 13$	54. $8 - 1$	55. $3 + 12$	56. $6 + 3$	57. $20 - 3$	58. $14 - 0$	59. $19 - 16$	60. $1 + 9$

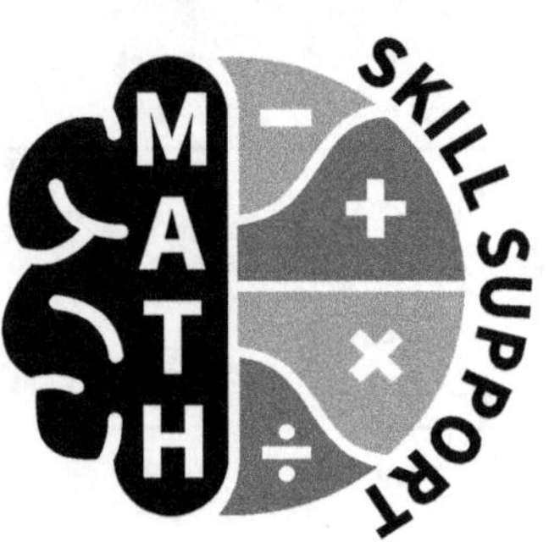

Name: ____________________ Date: ____________

				1. 2 + 7	2. 17 − 13	3. 6 − 5	4. 9 + 7
5. 1 + 10	6. 12 + 5	7. 11 − 4	8. 18 − 0	9. 20 − 7	10. 12 − 7	11. 14 + 1	12. 6 + 5
13. 13 − 5	14. 5 + 13	15. 12 + 6	16. 9 − 0	17. 6 + 0	18. 4 + 11	19. 19 − 2	20. 17 + 3
21. 5 + 10	22. 14 + 0	23. 7 − 2	24. 14 − 10	25. 16 − 7	26. 9 + 9	27. 18 − 6	28. 13 + 4
29. 1 + 17	30. 16 − 4	31. 11 + 4	32. 11 + 0	33. 8 + 8	34. 4 + 9	35. 15 − 10	36. 6 − 3
37. 18 − 9	38. 20 − 16	39. 14 + 6	40. 1 − 0	41. 13 − 9	42. 3 + 14	43. 7 + 6	44. 8 + 0
45. 9 − 8	46. 15 − 4	47. 10 + 2	48. 14 − 14	49. 16 − 6	50. 3 + 17	51. 7 + 1	52. 4 − 2
53. 2 + 12	54. 14 − 8	55. 4 + 15	56. 6 − 6	57. 4 − 4	58. 16 − 12	59. 19 − 0	60. 1 + 16

MIXED REVIEW: ADDING AND SUBTRACTING 0-20, SET E

Name: ____________________ Date: ____________________

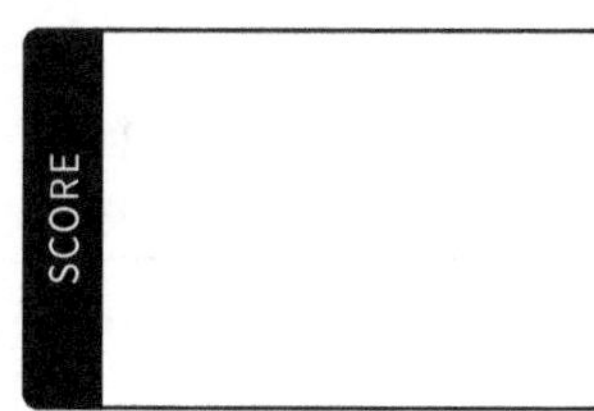

SCORE

				1) 14 -13	2) 12 - 5	3) 6 - 1	4) 19 -11
5) 17 -14	6) 16 -10	7) 20 + 0	8) 13 + 0	9) 13 -12	10) 3 +16	11) 10 - 9	12) 16 -11
13) 12 + 1	14) 17 - 0	15) 9 + 3	16) 4 +10	17) 5 + 0	18) 0 + 5	19) 12 - 0	20) 6 + 2
21) 0 +15	22) 20 - 8	23) 12 -12	24) 6 + 4	25) 18 -10	26) 1 +19	27) 11 - 0	28) 3 + 3
29) 11 + 1	30) 9 - 3	31) 16 - 8	32) 9 +10	33) 18 -14	34) 8 - 4	35) 2 + 4	36) 15 - 2
37) 5 +14	38) 9 - 7	39) 11 + 2	40) 8 - 3	41) 0 + 7	42) 5 + 1	43) 19 - 5	44) 20 -10
45) 10 + 7	46) 19 + 1	47) 20 -18	48) 3 + 4	49) 4 + 8	50) 8 + 7	51) 10 + 3	52) 19 -10
53) 15 + 5	54) 2 - 0	55) 10 + 4	56) 9 - 5	57) 20 -15	58) 17 - 6	59) 9 + 2	60) 7 +13

DAY 80

MIXED REVIEW: ADDING AND SUBTRACTING 0-20, SET F

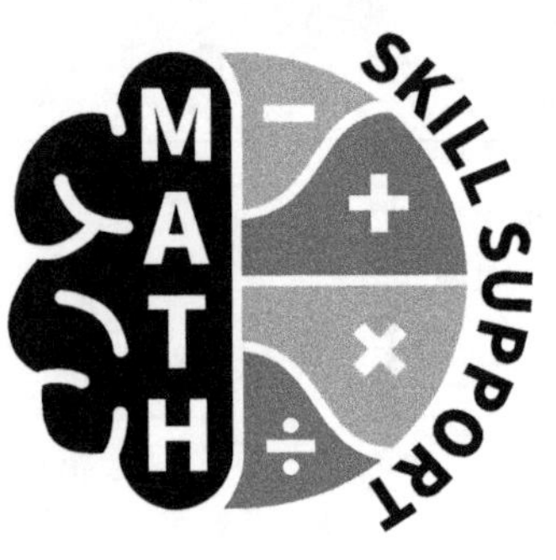

Name: ____________________ Date: ____________________

				1. 8 − 2	2. 20 − 5	3. 9 + 0	4. 13 − 0
5. 12 − 2	6. 20 − 6	7. 13 − 11	8. 8 + 1	9. 14 − 2	10. 0 + 17	11. 19 − 9	12. 12 + 3
13. 7 + 12	14. 15 − 15	15. 0 + 14	16. 7 + 8	17. 19 − 14	18. 2 + 16	19. 18 − 7	20. 1 + 12
21. 4 − 3	22. 10 − 3	23. 3 + 15	24. 7 + 0	25. 8 + 9	26. 15 + 4	27. 13 + 7	28. 14 + 4
29. 12 + 4	30. 7 − 6	31. 15 − 0	32. 3 + 7	33. 16 + 0	34. 16 − 5	35. 2 + 1	36. 2 + 0
37. 3 − 1	38. 15 − 1	39. 4 + 3	40. 20 − 14	41. 4 + 14	42. 9 − 9	43. 7 − 4	44. 10 − 5
45. 4 + 1	46. 8 + 6	47. 11 − 7	48. 15 − 13	49. 7 − 5	50. 17 + 1	51. 18 − 1	52. 2 + 18
53. 15 − 7	54. 20 − 13	55. 6 + 8	56. 6 + 7	57. 14 − 5	58. 1 + 8	59. 16 − 1	60. 2 + 9

Name: ____________________ Date: ____________

MIXED REVIEW: ADDING AND SUBTRACTING 0-20, SET G

#	Problem	#	Problem	#	Problem	#	Problem
1	8 + 10	2	5 + 8	3	16 − 15	4	18 − 5
5	6 − 2	6	10 − 2	7	17 − 3	8	0 + 8
9	13 + 6	10	19 − 17	11	1 + 13	12	19 − 19
13	3 + 9	14	9 − 2	15	3 + 0	16	1 + 2
17	0 + 3	18	7 + 7	19	10 − 0	20	16 + 4
21	9 + 8	22	3 + 8	23	8 + 4	24	10 − 8
25	16 − 16	26	4 + 7	27	0 + 6	28	2 + 5
29	20 − 11	30	19 + 0	31	17 − 5	32	11 − 9
33	2 + 15	34	13 − 3	35	16 − 2	36	9 + 4
37	15 − 9	38	8 − 5	39	5 + 4	40	1 + 1
41	15 − 12	42	6 − 4	43	2 + 6	44	6 + 12
45	1 + 0	46	9 + 1	47	12 − 9	48	10 + 0
49	12 + 8	50	18 − 18	51	5 − 3	52	10 + 9
53	4 + 13	54	18 − 11	55	10 − 10	56	20 − 20
57	11 − 11	58	20 − 4	59	16 − 9	60	17 − 9

MATH SKILL SUPPORT

Name: ____________________ Date: ____________

SCORE

1. 5 + 5
2. 12 − 2
3. 15 − 8
4. 18 − 12
5. 4 + 5
6. 10 + 9
7. 4 + 3
8. 17 + 0
9. 11 + 1
10. 20 − 18
11. 5 + 0
12. 17 − 9
13. 7 − 2
14. 5 + 2
15. 3 + 6
16. 13 − 9
17. 19 − 5
18. 15 + 4
19. 18 − 10
20. 2 + 14
21. 6 + 0
22. 14 − 7
23. 17 − 7
24. 20 − 15
25. 19 + 1
26. 4 + 2
27. 19 − 12
28. 6 + 6
29. 17 − 16
30. 14 − 4
31. 16 − 12
32. 2 + 11
33. 2 + 17
34. 1 + 2
35. 3 − 3
36. 0 + 18
37. 0 + 2
38. 15 − 7
39. 19 − 17
40. 14 + 2
41. 13 − 1
42. 10 − 8
43. 0 + 12
44. 10 − 9
45. 9 + 1
46. 19 − 0
47. 2 + 9
48. 12 − 0
49. 13 − 6
50. 20 − 20
51. 3 + 8
52. 6 − 6
53. 4 − 1
54. 3 + 1
55. 15 + 1
56. 18 − 0
57. 1 + 9
58. 13 − 3
59. 15 + 5
60. 11 + 0

MATH SKILL SUPPORT

Name: ____________________ Date: ____________________

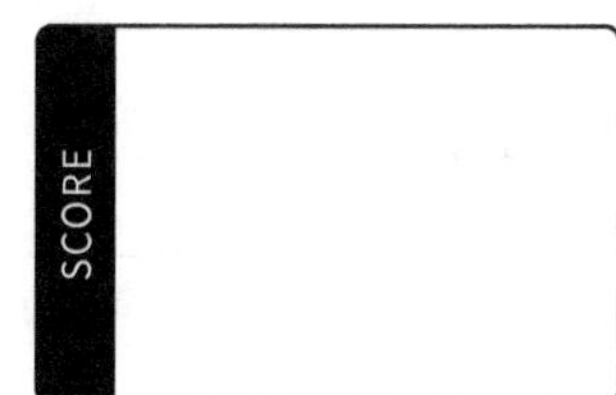
SCORE

DAY 83

MIXED REVIEW: ADDING AND SUBTRACTING 0-20, SET I

				1. 11 − 0	2. 7 + 12	3. 6 + 14	4. 17 − 3
5. 3 + 12	6. 20 − 8	7. 1 + 10	8. 4 + 9	9. 13 + 2	10. 20 − 14	11. 15 + 3	12. 3 − 2
13. 18 − 4	14. 7 + 8	15. 1 + 4	16. 18 − 5	17. 1 + 8	18. 20 − 16	19. 16 − 14	20. 18 + 2
21. 14 − 5	22. 8 + 1	23. 14 − 10	24. 10 − 10	25. 18 − 3	26. 11 + 4	27. 13 − 10	28. 1 + 13
29. 11 − 8	30. 6 + 4	31. 4 + 7	32. 5 + 8	33. 20 − 3	34. 2 − 0	35. 2 + 18	36. 15 − 1
37. 15 + 2	38. 17 − 8	39. 9 − 4	40. 18 − 18	41. 14 − 0	42. 3 + 11	43. 0 + 0	44. 7 + 7
45. 11 − 10	46. 0 + 17	47. 18 − 11	48. 6 + 5	49. 16 − 13	50. 19 − 15	51. 8 + 11	52. 11 + 7
53. 11 − 6	54. 12 − 9	55. 0 + 9	56. 5 + 15	57. 0 + 19	58. 13 − 13	59. 1 + 17	60. 12 − 11

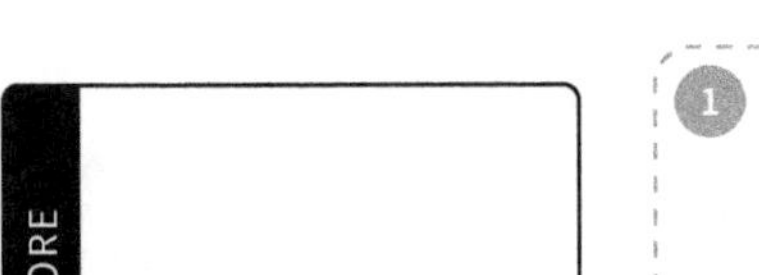

Name: ______________________ Date: ______________

SCORE

1. 18 − 1
2. 15 − 6
3. 0 + 6
4. 12 − 8
5. 15 − 14
6. 19 − 9
7. 3 + 10
8. 15 − 4
9. 0 + 20
10. 19 − 4
11. 1 − 1
12. 9 − 7
13. 11 − 3
14. 8 − 1
15. 2 − 2
16. 0 − 0
17. 10 − 7
18. 6 − 1
19. 9 − 2
20. 12 + 1
21. 1 + 0
22. 4 + 10
23. 5 + 1
24. 5 + 7
25. 4 + 0
26. 0 + 10
27. 16 + 0
28. 0 + 16
29. 3 + 15
30. 13 − 7
31. 10 + 0
32. 3 + 3
33. 14 − 13
34. 13 + 1
35. 14 + 6
36. 13 + 6
37. 1 + 18
38. 10 − 5
39. 16 − 11
40. 8 + 7
41. 5 + 11
42. 6 + 7
43. 7 − 7
44. 3 − 0
45. 10 − 4
46. 4 − 3
47. 10 − 3
48. 4 + 8
49. 9 + 0
50. 11 − 4
51. 15 − 2
52. 10 + 1
53. 17 − 5
54. 7 + 1
55. 13 + 4
56. 16 − 5
57. 8 − 8
58. 10 + 8
59. 11 + 2
60. 8 + 3

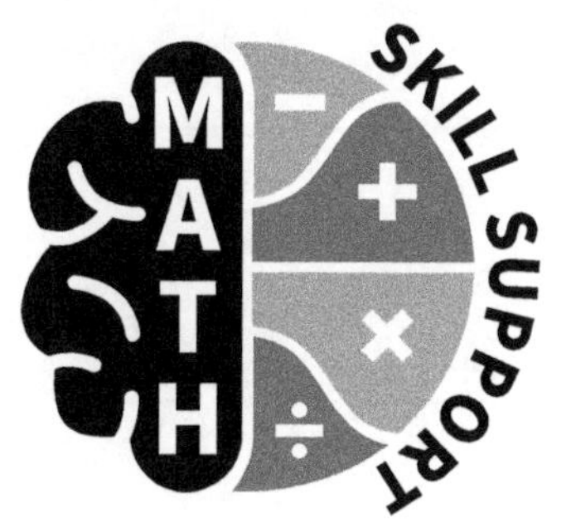

Name: ____________________ Date: ____________

DAY 85

MIXED REVIEW: ADDING AND SUBTRACTING 0-20, SET K

				(1) 8 − 5	(2) 12 + 0	(3) 10 + 5	(4) 17 − 15
(5) 6 + 2	(6) 7 + 11	(7) 7 + 3	(8) 4 + 1	(9) 15 − 11	(10) 19 − 10	(11) 6 − 3	(12) 9 − 0
(13) 14 + 5	(14) 20 − 17	(15) 6 + 9	(16) 8 − 3	(17) 2 + 15	(18) 16 + 1	(19) 5 + 6	(20) 6 + 3
(21) 14 − 8	(22) 10 − 6	(23) 9 + 3	(24) 14 + 4	(25) 17 − 4	(26) 19 − 14	(27) 4 − 2	(28) 1 + 14
(29) 11 − 5	(30) 3 + 7	(31) 8 + 8	(32) 9 + 2	(33) 16 − 10	(34) 16 − 3	(35) 5 − 3	(36) 18 − 7
(37) 5 + 3	(38) 10 − 1	(39) 8 + 5	(40) 1 + 12	(41) 7 + 4	(42) 6 + 8	(43) 2 + 3	(44) 19 − 13
(45) 18 − 16	(46) 18 − 15	(47) 16 − 9	(48) 7 − 4	(49) 7 + 5	(50) 4 + 16	(51) 20 − 10	(52) 17 − 14
(53) 8 + 12	(54) 12 + 3	(55) 5 + 12	(56) 7 − 3	(57) 10 + 6	(58) 16 − 8	(59) 15 − 9	(60) 13 − 4

MATH SKILL SUPPORT

Name: ____________________ Date: ____________

SCORE

#	Problem	#	Problem	#	Problem	#	Problem
1	9 − 3	2	16 − 0	3	0 + 14	4	17 − 12
5	9 − 5	6	6 + 10	7	2 + 4	8	18 − 14
9	11 − 11	10	12 + 7	11	1 + 16	12	3 + 5
13	18 − 6	14	5 + 9	15	14 − 6	16	10 − 0
17	14 − 3	18	5 + 4	19	19 − 18	20	11 − 9
21	3 + 16	22	0 + 4	23	16 − 4	24	18 − 2
25	20 − 4	26	11 − 2	27	9 + 7	28	11 + 9
29	15 − 5	30	14 − 9	31	8 − 4	32	9 + 9
33	2 + 10	34	20 − 9	35	18 − 8	36	7 − 0
37	16 + 2	38	7 + 2	39	3 + 4	40	9 + 5
41	9 + 6	42	13 − 0	43	2 + 0	44	15 − 15
45	7 + 13	46	12 + 2	47	5 − 0	48	3 + 17
49	2 − 1	50	12 − 1	51	2 + 6	52	20 − 1
53	2 + 12	54	4 + 14	55	11 + 8	56	10 + 2
57	10 + 4	58	1 + 5	59	1 − 0	60	20 − 12

MATH SKILL SUPPORT

Name: ______________________ Date: ______________

SCORE

DAY 87

MIXED REVIEW: ADDING AND SUBTRACTING 0-20, SET M

				(1) 6 + 11	(2) 6 + 13	(3) 2 + 2	(4) 1 + 7
(5) 7 − 1	(6) 13 + 0	(7) 16 + 3	(8) 0 + 1	(9) 14 + 0	(10) 19 − 6	(11) 19 − 11	(12) 20 − 13
(13) 13 − 5	(14) 3 + 0	(15) 7 + 10	(16) 2 + 1	(17) 5 − 5	(18) 10 − 2	(19) 7 − 5	(20) 1 + 15
(21) 1 + 1	(22) 6 − 2	(23) 2 + 16	(24) 17 − 1	(25) 1 + 11	(26) 15 + 0	(27) 8 − 6	(28) 9 + 10
(29) 16 − 7	(30) 12 + 5	(31) 18 + 1	(32) 18 + 0	(33) 5 + 14	(34) 3 + 2	(35) 8 + 6	(36) 1 + 3
(37) 19 − 3	(38) 14 − 2	(39) 15 − 13	(40) 7 − 6	(41) 14 − 1	(42) 15 − 10	(43) 19 − 2	(44) 13 + 5
(45) 8 − 2	(46) 12 − 3	(47) 1 + 6	(48) 17 + 3	(49) 20 − 5	(50) 17 − 0	(51) 5 + 13	(52) 9 + 4
(53) 9 − 8	(54) 8 + 0	(55) 3 − 1	(56) 17 − 10	(57) 13 − 8	(58) 9 − 9	(59) 9 − 1	(60) 5 − 2

DAY 88

MIXED REVIEW: ADDING AND SUBTRACTING 0-20, SET N

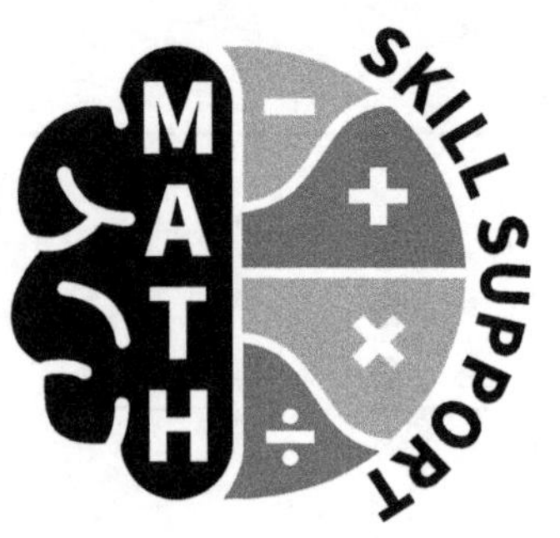

Name: ______________________ Date: ______________

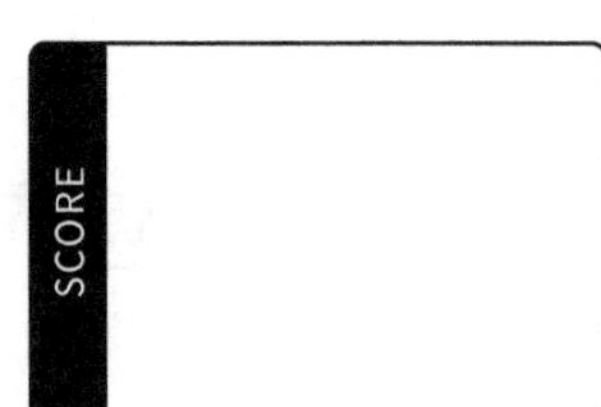

				① 17 + 2	② 6 − 4	③ 16 − 1	④ 0 +13
⑤ 3 +13	⑥ 14 −14	⑦ 8 − 0	⑧ 8 + 4	⑨ 13 − 2	⑩ 17 −13	⑪ 9 +11	⑫ 16 + 4
⑬ 19 − 1	⑭ 4 + 6	⑮ 12 − 7	⑯ 19 −16	⑰ 1 +19	⑱ 8 + 2	⑲ 15 −12	⑳ 20 − 2
21: 6 + 1	22: 17 − 6	23: 17 −11	24: 11 + 6	25: 2 + 7	26: 10 + 3	27: 16 − 2	28: 19 + 0
29: 12 + 8	30: 2 + 8	31: 17 − 2	32: 0 + 7	33: 17 + 1	34: 16 −16	35: 6 − 0	36: 3 +14
37: 12 − 5	38: 8 − 7	39: 12 + 4	40: 18 − 9	41: 19 − 7	42: 19 − 8	43: 4 − 4	44: 5 +10
45: 9 + 8	46: 0 + 8	47: 4 +12	48: 4 +15	49: 18 −13	50: 20 + 0	51: 7 + 6	52: 20 −19
53: 6 +12	54: 14 −12	55: 12 −10	56: 20 − 7	57: 9 − 6	58: 2 +13	59: 0 +15	60: 13 −11

Name: ____________________ Date: ____________

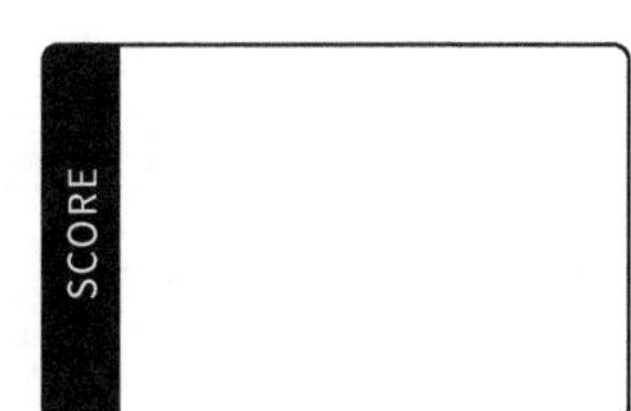

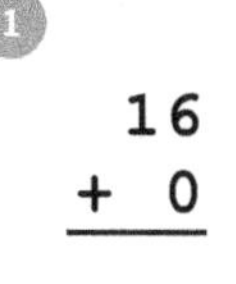

1) 16 + 0	2) 11 − 11	3) 12 − 1	4) 13 − 8

5) 11 + 0	6) 11 + 2	7) 13 − 6	8) 9 + 10	9) 16 − 1	10) 18 − 1	11) 14 − 12	12) 11 − 5
13) 0 + 16	14) 18 − 14	15) 1 + 18	16) 10 − 9	17) 11 − 1	18) 2 + 13	19) 11 + 4	20) 8 − 1
21) 6 − 6	22) 2 + 0	23) 2 + 3	24) 8 − 7	25) 19 + 1	26) 16 + 4	27) 15 − 12	28) 10 − 7
29) 12 − 11	30) 11 − 2	31) 15 + 4	32) 9 + 5	33) 12 + 3	34) 20 − 20	35) 9 − 3	36) 0 + 8
37) 0 + 14	38) 15 − 14	39) 12 + 1	40) 20 − 16	41) 4 + 8	42) 8 + 6	43) 10 − 3	44) 10 + 4
45) 1 + 14	46) 9 + 7	47) 20 − 4	48) 18 − 15	49) 7 + 13	50) 0 + 6	51) 8 − 4	52) 3 + 14
53) 4 + 1	54) 4 + 13	55) 7 + 7	56) 20 − 3	57) 3 − 0	58) 18 − 7	59) 12 − 12	60) 20 + 0

MIXED REVIEW: ADDING AND SUBTRACTING 0-20, SET P

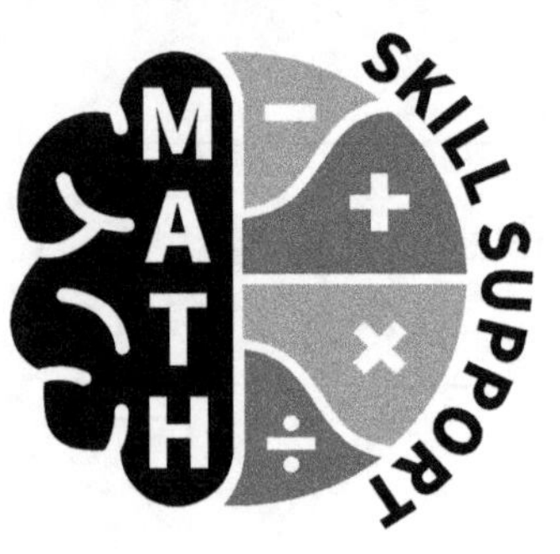

Name: ____________________ Date: ____________

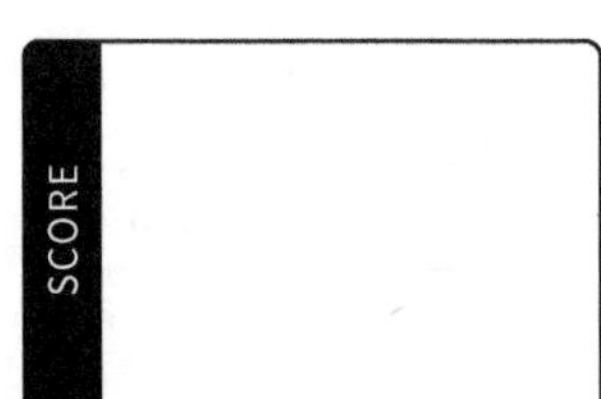

				(1) 1 + 7	(2) 11 − 0	(3) 5 + 9	(4) 15 − 9
(5) 6 + 13	(6) 13 − 12	(7) 7 + 9	(8) 16 − 0	(9) 16 − 2	(10) 11 + 5	(11) 13 − 10	(12) 19 − 6
(13) 9 − 2	(14) 20 − 14	(15) 6 − 1	(16) 10 + 8	(17) 6 + 9	(18) 7 − 0	(19) 5 − 4	(20) 15 − 10
(21) 12 + 7	(22) 7 + 0	(23) 13 + 4	(24) 4 − 4	(25) 5 − 5	(26) 9 − 4	(27) 13 − 1	(28) 3 + 3
(29) 4 + 9	(30) 6 + 12	(31) 1 + 15	(32) 4 + 6	(33) 14 + 6	(34) 14 + 4	(35) 16 − 16	(36) 7 − 3
(37) 11 + 8	(38) 20 − 13	(39) 19 − 13	(40) 14 − 5	(41) 17 − 14	(42) 8 + 10	(43) 7 − 5	(44) 11 − 3
(45) 4 + 5	(46) 14 + 2	(47) 1 + 10	(48) 2 + 16	(49) 6 + 3	(50) 7 + 3	(51) 12 − 2	(52) 12 − 6
(53) 4 + 0	(54) 2 + 4	(55) 13 − 5	(56) 9 + 4	(57) 14 + 5	(58) 2 + 9	(59) 19 − 15	(60) 16 − 13

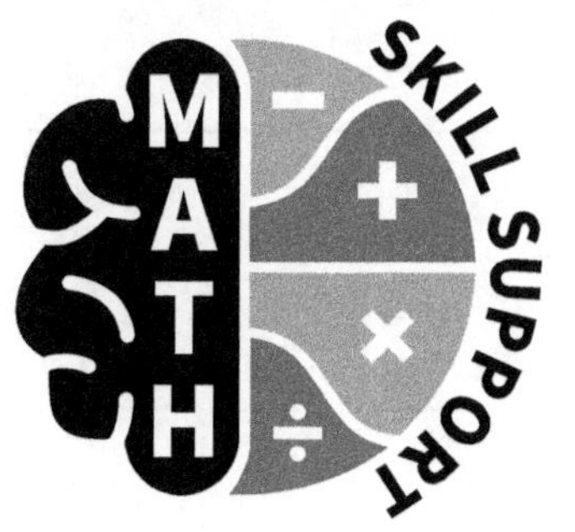

Name: ____________________ Date: ____________________

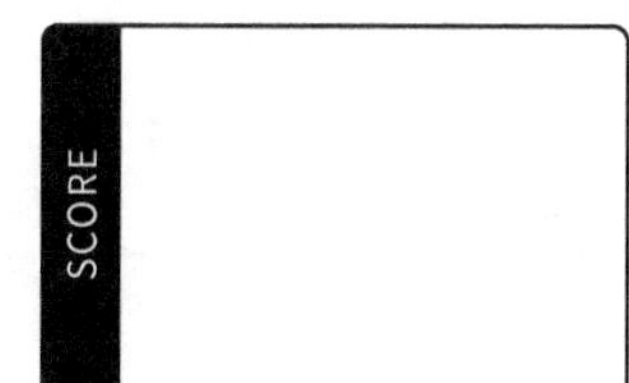

MIXED REVIEW: ADDING AND SUBTRACTING 0-20, SET Q

1. $2 + 18$	2. $6 + 2$	3. $5 + 2$	4. $12 - 0$

5. $12 + 4$	6. $14 - 11$	7. $7 - 6$	8. $13 - 11$	9. $3 + 2$	10. $20 - 18$	11. $13 - 0$	12. $12 - 4$
13. $9 + 0$	14. $9 + 8$	15. $5 + 11$	16. $0 + 17$	17. $14 - 0$	18. $11 + 7$	19. $8 - 8$	20. $20 - 7$
21. $10 + 3$	22. $11 + 6$	23. $10 - 10$	24. $17 - 6$	25. $2 + 17$	26. $13 + 2$	27. $10 - 0$	28. $14 + 1$
29. $11 - 9$	30. $16 - 9$	31. $15 - 3$	32. $11 - 6$	33. $11 + 3$	34. $5 + 6$	35. $1 + 0$	36. $19 - 7$
37. $8 + 1$	38. $7 + 6$	39. $13 + 6$	40. $1 + 16$	41. $4 + 14$	42. $1 + 11$	43. $7 - 4$	44. $18 - 16$
45. $18 - 2$	46. $2 + 11$	47. $17 - 9$	48. $4 + 3$	49. $9 - 5$	50. $6 + 7$	51. $18 - 9$	52. $17 + 1$
53. $16 - 7$	54. $19 - 8$	55. $9 + 3$	56. $20 - 5$	57. $15 - 0$	58. $16 + 3$	59. $14 - 13$	60. $8 - 5$

MIXED REVIEW: ADDING AND SUBTRACTING 0-20, SET R

MATH SKILL SUPPORT

Name: ____________________ Date: ____________________

1. 0 + 13	2. 18 - 6	3. 16 - 5	4. 10 + 7

5. 14 - 3	6. 0 + 7	7. 3 + 5	8. 12 + 2	9. 14 - 14	10. 2 - 2	11. 15 - 13	12. 3 - 2
13. 2 + 14	14. 12 - 3	15. 12 - 7	16. 0 + 9	17. 19 - 5	18. 19 - 0	19. 20 - 17	20. 8 + 4
21. 0 + 15	22. 13 - 13	23. 16 - 10	24. 0 + 3	25. 15 - 8	26. 3 + 11	27. 0 + 0	28. 13 + 1
29. 13 - 9	30. 2 - 1	31. 5 + 7	32. 9 - 7	33. 7 - 2	34. 0 + 12	35. 1 + 5	36. 10 - 5
37. 1 + 13	38. 8 - 3	39. 0 + 2	40. 0 + 10	41. 18 + 0	42. 11 - 8	43. 18 - 8	44. 18 - 5
45. 17 - 11	46. 3 + 15	47. 15 - 5	48. 8 + 8	49. 14 - 8	50. 3 + 13	51. 2 + 15	52. 7 + 1
53. 1 + 19	54. 20 - 6	55. 15 - 15	56. 15 + 3	57. 19 - 10	58. 7 + 2	59. 3 + 6	60. 3 + 8

MATH SKILL SUPPORT

Name: ______________________ Date: ______________

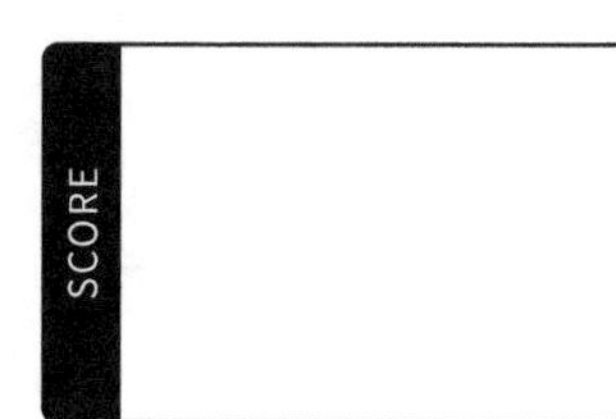

SCORE

MIXED REVIEW: ADDING AND SUBTRACTING 0-20, SET S

				1. 14 - 2	2. 0 +18	3. 15 -11	4. 0 +19
5. 8 + 2	6. 16 + 2	7. 17 + 3	8. 13 + 5	9. 16 -14	10. 13 - 3	11. 17 -16	12. 19 - 1
13. 2 + 2	14. 13 - 7	15. 18 + 1	16. 14 + 0	17. 5 - 3	18. 9 + 9	19. 5 +10	20. 18 -17
21. 5 + 3	22. 20 -11	23. 17 + 2	24. 12 + 8	25. 14 - 4	26. 16 - 4	27. 19 - 3	28. 17 + 0
29. 17 - 8	30. 10 + 9	31. 1 + 3	32. 16 - 3	33. 10 + 6	34. 2 + 8	35. 0 +20	36. 17 -17
37. 8 - 6	38. 8 + 3	39. 2 +10	40. 9 - 6	41. 3 - 3	42. 17 - 4	43. 5 +12	44. 6 - 0
45. 9 + 1	46. 11 - 7	47. 11 - 4	48. 12 - 9	49. 4 +12	50. 4 + 2	51. 15 - 1	52. 1 + 2
53. 16 + 1	54. 8 + 7	55. 16 -12	56. 20 - 2	57. 9 - 1	58. 9 - 0	59. 15 - 4	60. 6 + 4

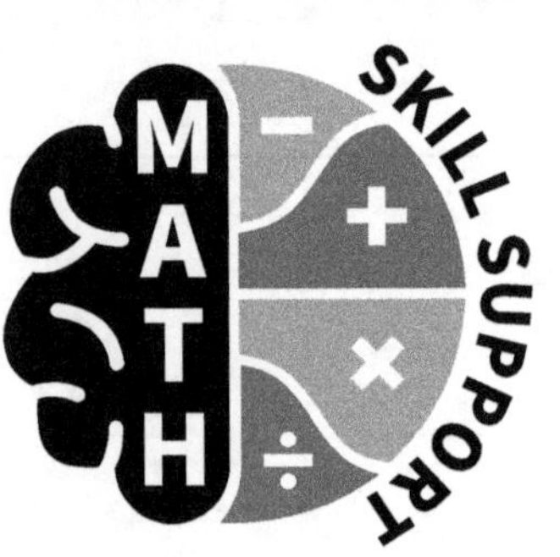

Name: ____________________ Date: ____________

				1) 5 − 2	2) 2 + 7	3) 7 + 4	4) 10 − 6
5) 1 + 4	6) 4 − 0	7) 20 − 8	8) 7 + 5	9) 7 +12	10) 6 − 3	11) 15 − 2	12) 4 +11
13) 3 +16	14) 19 −11	15) 16 − 8	16) 12 + 0	17) 20 −15	18) 4 + 4	19) 18 − 0	20) 5 + 5
21) 8 − 2	22) 1 + 6	23) 3 +12	24) 15 + 2	25) 0 + 4	26) 1 − 0	27) 10 − 8	28) 4 − 3
29) 17 − 7	30) 3 + 7	31) 3 + 0	32) 6 − 5	33) 20 − 1	34) 1 +17	35) 3 +17	36) 9 + 6
37) 1 − 1	38) 13 − 2	39) 16 −15	40) 6 + 8	41) 10 +10	42) 20 −12	43) 8 − 0	44) 8 + 9
45) 2 + 5	46) 15 + 1	47) 8 + 0	48) 14 − 9	49) 17 −15	50) 12 − 5	51) 1 + 9	52) 5 + 8
53) 3 +10	54) 3 + 9	55) 19 − 4	56) 7 − 7	57) 15 − 6	58) 5 + 1	59) 3 − 1	60) 11 −10

MATH SKILL SUPPORT

Name: ____________________ Date: ____________

1. $5 - 0$
2. $17 - 5$
3. $17 - 3$
4. $18 - 4$
5. $19 + 0$
6. $8 + 12$
7. $4 + 7$
8. $15 + 5$
9. $9 - 8$
10. $15 - 7$
11. $20 - 10$
12. $1 + 1$
13. $11 + 9$
14. $14 - 7$
15. $6 + 6$
16. $13 + 3$
17. $14 - 1$
18. $4 - 2$
19. $2 + 12$
20. $10 - 2$
21. $17 - 0$
22. $6 + 10$
23. $2 - 0$
24. $0 + 11$
25. $0 - 0$
26. $4 + 15$
27. $19 - 2$
28. $10 - 4$
29. $5 - 1$
30. $12 + 5$
31. $18 - 11$
32. $3 + 1$
33. $7 - 1$
34. $18 - 3$
35. $17 - 13$
36. $0 + 1$
37. $10 + 2$
38. $14 + 3$
39. $12 + 6$
40. $19 - 12$
41. $6 + 14$
42. $18 - 18$
43. $14 - 10$
44. $7 + 8$
45. $18 - 13$
46. $8 + 5$
47. $5 + 15$
48. $14 - 6$
49. $3 + 4$
50. $16 - 6$
51. $7 + 11$
52. $10 + 1$
53. $0 + 5$
54. $10 - 1$
55. $5 + 14$
56. $4 + 10$
57. $6 - 2$
58. $6 + 5$
59. $5 + 0$
60. $9 - 9$

MATH SKILL SUPPORT

Name: ____________________ Date: ____________________

SCORE

				1. 9 + 0 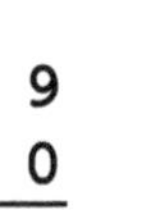	2. 2 + 3 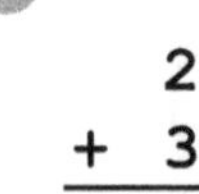	3. 14 − 2	4. 0 + 0
5. 18 + 1	6. 12 − 4	7. 18 − 17	8. 6 − 6	9. 5 + 0	10. 13 + 1	11. 19 + 1	12. 10 − 8
13. 13 + 3	14. 1 + 2	15. 15 + 1	16. 11 − 10	17. 5 + 8	18. 18 − 15	19. 18 − 8	20. 5 + 3
21. 4 + 13	22. 13 − 6	23. 15 + 0	24. 15 − 12	25. 3 + 14	26. 0 + 1	27. 3 + 10	28. 10 − 3
29. 8 + 5	30. 1 + 15	31. 16 − 6	32. 0 + 13	33. 20 − 3	34. 7 − 6	35. 2 + 10	36. 17 − 4
37. 12 − 6	38. 20 − 10	39. 8 + 1	40. 1 + 13	41. 14 − 0	42. 13 − 4	43. 9 + 1	44. 2 + 8
45. 20 − 19	46. 2 + 12	47. 8 + 3	48. 15 − 2	49. 18 − 12	50. 1 + 12	51. 17 − 6	52. 15 − 6
53. 10 − 1	54. 20 − 4	55. 19 − 16	56. 0 + 6	57. 7 − 5	58. 15 − 14	59. 1 + 4	60. 20 − 1

SKILL SUPPORT MATH

Name: ____________________ Date: ____________________

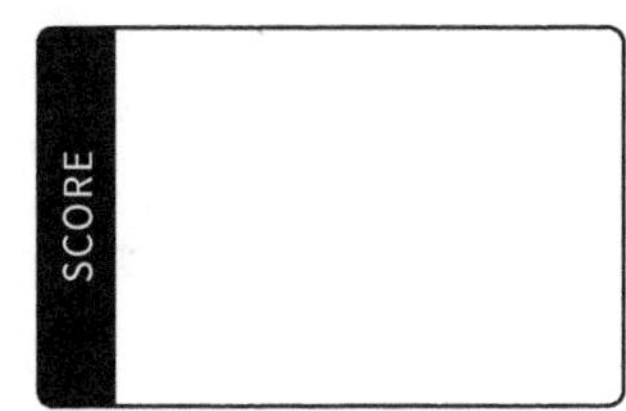

MIXED REVIEW: ADDING AND SUBTRACTING 0-20, SET W

				(1) 11 + 3	(2) 8 + 12	(3) 14 − 4	(4) 17 − 11
(5) 12 + 6	(6) 6 + 13	(7) 12 + 0	(8) 12 + 2	(9) 10 + 0	(10) 20 − 17	(11) 16 − 8	(12) 16 − 4
(13) 3 − 1	(14) 18 − 14	(15) 12 + 8	(16) 0 + 16	(17) 9 + 11	(18) 5 − 0	(19) 17 − 8	(20) 14 + 1
(21) 3 + 6	(22) 13 − 11	(23) 4 + 14	(24) 1 + 19	(25) 14 − 14	(26) 14 + 3	(27) 14 − 6	(28) 6 + 12
(29) 17 − 7	(30) 3 + 2	(31) 2 + 2	(32) 20 − 5	(33) 13 − 12	(34) 17 − 1	(35) 19 − 14	(36) 10 − 9
(37) 4 + 2	(38) 2 − 1	(39) 1 + 3	(40) 9 − 9	(41) 13 − 5	(42) 4 + 4	(43) 19 − 3	(44) 15 + 2
(45) 3 − 3	(46) 16 − 7	(47) 0 + 8	(48) 12 + 5	(49) 8 − 6	(50) 10 + 1	(51) 6 + 10	(52) 16 + 4
(53) 4 − 1	(54) 4 + 6	(55) 15 − 3	(56) 17 − 16	(57) 0 + 10	(58) 10 − 5	(59) 2 + 14	(60) 13 − 3

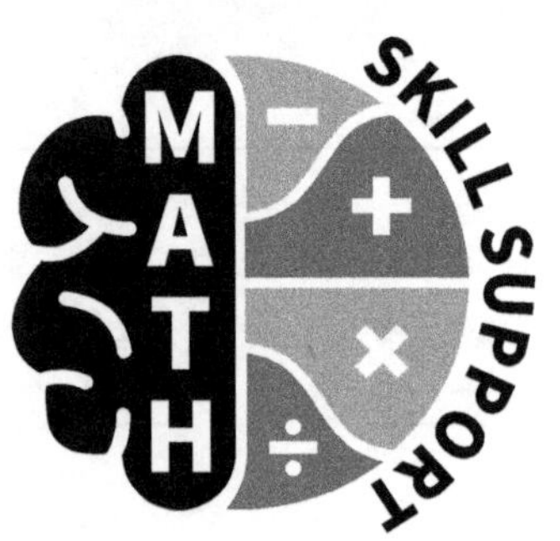

Name: ______________________ Date: ______________

				1. $6 + 5$	2. $7 + 11$	3. $6 - 5$	4. $19 - 17$
5. $14 + 2$	6. $14 - 12$	7. $11 + 6$	8. $17 - 17$	9. $10 - 10$	10. $5 + 12$	11. $11 - 2$	12. $4 + 8$
13. $20 - 15$	14. $0 - 0$	15. $0 + 19$	16. $8 + 4$	17. $8 - 1$	18. $19 - 11$	19. $7 + 0$	20. $1 + 18$
21. $4 + 11$	22. $11 - 1$	23. $5 + 5$	24. $14 - 7$	25. $7 - 2$	26. $2 + 7$	27. $12 + 7$	28. $7 + 12$
29. $11 + 0$	30. $2 + 9$	31. $0 + 7$	32. $13 - 0$	33. $4 - 4$	34. $4 + 5$	35. $18 - 2$	36. $1 + 5$
37. $6 + 14$	38. $10 + 4$	39. $15 - 8$	40. $14 - 5$	41. $16 - 11$	42. $12 - 3$	43. $6 + 2$	44. $4 - 0$
45. $16 + 0$	46. $5 + 7$	47. $20 - 16$	48. $2 + 17$	49. $16 - 13$	50. $17 - 14$	51. $8 - 8$	52. $4 + 0$
53. $3 - 2$	54. $16 - 3$	55. $1 + 1$	56. $7 - 4$	57. $10 - 7$	58. $10 + 3$	59. $8 + 2$	60. $12 - 1$

Name: ______________________ Date: ______________

MIXED REVIEW: ADDING AND SUBTRACTING 0-20, SET Y

				(1) 18 - 1	(2) 11 + 2	(3) 14 + 5	(4) 19 - 5
(5) 18 -10	(6) 20 - 9	(7) 11 - 8	(8) 0 + 3	(9) 10 - 2	(10) 18 -13	(11) 8 +11	(12) 19 + 0
(13) 8 + 6	(14) 9 - 5	(15) 1 +17	(16) 12 -12	(17) 19 -10	(18) 4 + 9	(19) 17 - 2	(20) 18 - 0
(21) 13 - 8	(22) 20 -14	(23) 3 + 9	(24) 9 + 2	(25) 14 + 6	(26) 6 + 7	(27) 13 -10	(28) 0 + 4
(29) 2 +11	(30) 6 + 8	(31) 15 - 7	(32) 6 - 4	(33) 20 - 6	(34) 19 - 4	(35) 5 +14	(36) 11 + 8
(37) 0 + 5	(38) 3 +17	(39) 6 + 9	(40) 11 -11	(41) 19 -19	(42) 16 -16	(43) 3 + 4	(44) 18 - 3
(45) 18 -16	(46) 9 + 8	(47) 11 - 4	(48) 17 - 0	(49) 4 +10	(50) 0 +15	(51) 9 - 6	(52) 17 - 9
(53) 7 + 3	(54) 13 - 7	(55) 2 +18	(56) 17 + 2	(57) 5 +15	(58) 11 - 3	(59) 7 + 7	(60) 7 +13

DAY 100

MIXED REVIEW: ADDING AND SUBTRACTING 0-20, SET Z

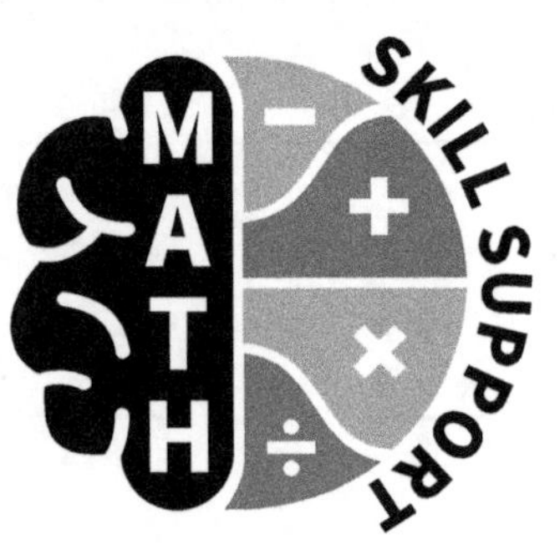

Name: ______________________ Date: ______________

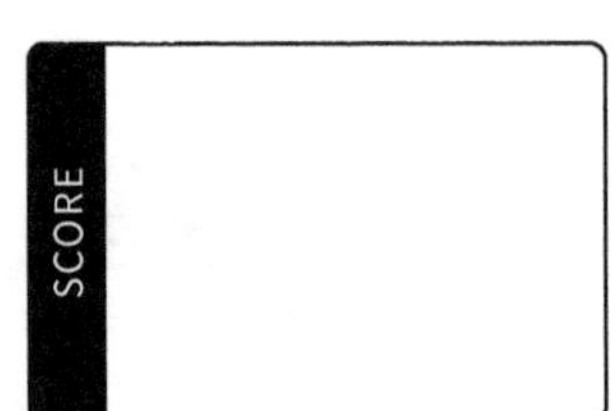

				1. 8 + 8	2. 1 + 6	3. 0 + 9	4. 9 + 7
5. 12 − 7	6. 12 − 5	7. 4 +15	8. 5 + 2	9. 10 + 6	10. 10 + 5	11. 16 −14	12. 6 + 1
13. 2 +13	14. 19 − 8	15. 4 − 3	16. 7 + 8	17. 11 − 6	18. 12 − 2	19. 20 −12	20. 12 + 1
21. 11 + 4	22. 17 + 3	23. 17 −10	24. 2 +15	25. 14 −11	26. 3 + 3	27. 9 + 9	28. 5 − 2
29. 7 + 6	30. 6 + 0	31. 16 −15	32. 16 − 0	33. 3 +16	34. 15 − 9	35. 11 − 9	36. 19 −18
37. 5 +13	38. 0 +12	39. 12 −10	40. 17 − 5	41. 7 + 2	42. 8 − 0	43. 19 − 9	44. 5 + 9
45. 8 + 7	46. 15 −11	47. 8 − 2	48. 4 +12	49. 4 + 7	50. 0 + 2	51. 12 − 9	52. 17 −15
53. 17 −13	54. 18 −18	55. 18 + 2	56. 13 − 2	57. 7 − 3	58. 17 −12	59. 9 +10	60. 15 −10

ANSWER KEY

Day 1	Day 2	Day 3	Day 4	Day 5	Day 6	Day 7	Day 8	Day 9	Day 10
1) 5	1) 16	1) 18	1) 18	1) 16	1) 17	1) 0	1) 7	1) 8	1) 4
2) 10	2) 9	2) 10	2) 20	2) 11	2) 16	2) 1	2) 8	2) 9	2) 10
3) 19	3) 10	3) 4	3) 11	3) 19	3) 20	3) 2	3) 1	3) 5	3) 3
4) 11	4) 14	4) 8	4) 5	4) 11	4) 13	4) 3	4) 3	4) 6	4) 9
5) 7	5) 9	5) 1	5) 15	5) 12	5) 12	5) 4	5) 10	5) 4	5) 1
6) 1	6) 4	6) 10	6) 12	6) 20	6) 11	6) 5	6) 10	6) 6	6) 6
7) 15	7) 18	7) 17	7) 12	7) 18	7) 18	7) 6	7) 4	7) 8	7) 2
8) 6	8) 20	8) 5	8) 14	8) 16	8) 17	8) 7	8) 3	8) 5	8) 9
9) 4	9) 8	9) 2	9) 1	9) 10	9) 13	9) 8	9) 7	9) 8	9) 9
10) 13	10) 4	10) 19	10) 7	10) 14	10) 10	10) 9	10) 6	10) 3	10) 3
11) 2	11) 0	11) 9	11) 19	11) 15	11) 14	11) 10	11) 6	11) 5	11) 9
12) 2	12) 14	12) 11	12) 16	12) 17	12) 10	12) 1	12) 9	12) 9	12) 5
13) 20	13) 10	13) 12	13) 16	13) 18	13) 15	13) 2	13) 5	13) 8	13) 8
14) 7	14) 3	14) 13	14) 20	14) 15	14) 19	14) 3	14) 10	14) 2	14) 6
15) 13	15) 6	15) 14	15) 3	15) 20	15) 15	15) 4	15) 5	15) 9	15) 6
16) 20	16) 6	16) 14	16) 8	16) 12	16) 20	16) 5	16) 4	16) 2	16) 7
17) 3	17) 17	17) 17	17) 17	17) 19	17) 19	17) 6	17) 10	17) 7	17) 8
18) 3	18) 20	18) 7	18) 15	18) 13	18) 16	18) 7	18) 9	18) 6	18) 5
19) 12	19) 12	19) 3	19) 4	19) 13	19) 14	19) 8	19) 8	19) 10	19) 1
20) 18	20) 5	20) 1	20) 3	20) 14	20) 18	20) 9	20) 9	20) 10	20) 9
21) 9	21) 13	21) 7	21) 6	21) 17	21) 12	21) 10	21) 0	21) 7	21) 8
22) 12	22) 13	22) 11	22) 5	22) 10	22) 11	22) 2	22) 8	22) 9	22) 10
23) 8	23) 1	23) 8	23) 13	23) 20	23) 18	23) 3	23) 9	23) 7	23) 6
24) 14	24) 15	24) 19	24) 2	24) 19	24) 11	24) 4	24) 9	24) 4	24) 6
25) 16	25) 5	25) 4	25) 10	25) 11	25) 18	25) 5	25) 8	25) 7	25) 8
26) 10	26) 3	26) 12	26) 6	26) 16	26) 16	26) 6	26) 6	26) 10	26) 5
27) 16	27) 16	27) 6	27) 7	27) 14	27) 17	27) 7	27) 7	27) 9	27) 8
28) 4	28) 18	28) 5	28) 11	28) 15	28) 13	28) 8	28) 8	28) 8	28) 7
29) 17	29) 8	29) 16	29) 9	29) 12	29) 20	29) 9	29) 3	29) 10	29) 4
30) 19	30) 2	30) 15	30) 8	30) 19	30) 16	30) 10	30) 3	30) 8	30) 7
31) 15	31) 19	31) 20	31) 1	31) 10	31) 15	31) 3	31) 9	31) 0	31) 7
32) 9	32) 1	32) 15	32) 9	32) 17	32) 14	32) 4	32) 5	32) 10	32) 0
33) 6	33) 2	33) 13	33) 10	33) 15	33) 12	33) 5	33) 2	33) 8	33) 4
34) 8	34) 17	34) 18	34) 18	34) 11	34) 20	34) 6	34) 10	34) 8	34) 2
35) 11	35) 15	35) 3	35) 13	35) 16	35) 19	35) 7	35) 2	35) 10	35) 8
36) 5	36) 11	36) 20	36) 17	36) 13	36) 11	36) 8	36) 10	36) 3	36) 7
37) 18	37) 12	37) 9	37) 14	37) 18	37) 10	37) 9	37) 5	37) 6	37) 8
38) 17	38) 7	38) 6	38) 19	38) 20	38) 19	38) 10	38) 2	38) 9	38) 10
39) 0	39) 7	39) 16	39) 4	39) 18	39) 15	39) 4	39) 4	39) 9	39) 3
40) 1	40) 11	40) 13	40) 3	40) 10	40) 13	40) 5	40) 4	40) 9	40) 10
41) 14	41) 19	41) 16	41) 13	41) 13	41) 12	41) 6	41) 9	41) 5	41) 9
42) 10	42) 12	42) 7	42) 16	42) 12	42) 14	42) 7	42) 4	42) 8	42) 10
43) 7	43) 12	43) 4	43) 14	43) 14	43) 10	43) 8	43) 7	43) 7	43) 9
44) 13	44) 4	44) 10	44) 20	44) 17	44) 17	44) 9	44) 8	44) 10	44) 10
45) 7	45) 6	45) 18	45) 11	45) 15	45) 12	45) 10	45) 8	45) 10	45) 10
46) 3	46) 7	46) 16	46) 4	46) 19	46) 18	46) 5	46) 7	46) 9	46) 7
47) 13	47) 3	47) 1	47) 1	47) 11	47) 14	47) 6	47) 10	47) 7	47) 6
48) 8	48) 11	48) 5	48) 16	48) 13	48) 13	48) 7	48) 7	48) 6	48) 9
49) 20	49) 11	49) 4	49) 5	49) 20	49) 17	49) 8	49) 10	49) 4	49) 4
50) 17	50) 8	50) 17	50) 10	50) 14	50) 17	50) 9	50) 9	50) 7	50) 5
51) 3	51) 4	51) 8	51) 7	51) 12	51) 10	51) 10	51) 7	51) 3	51) 6
52) 16	52) 15	52) 5	52) 19	52) 18	52) 20	52) 6	52) 7	52) 9	52) 7
53) 6	53) 14	53) 11	53) 10	53) 19	53) 10	53) 7	53) 9	53) 10	53) 9
54) 20	54) 14	54) 3	54) 9	54) 13	54) 13	54) 8	54) 10	54) 10	54) 3
55) 1	55) 2	55) 19	55) 9	55) 14	55) 15	55) 9	55) 10	55) 7	55) 2
56) 5	56) 18	56) 11	56) 11	56) 15	56) 12	56) 10	56) 8	56) 6	56) 10
57) 0	57) 10	57) 2	57) 8	57) 11	57) 19	57) 7	57) 1	57) 5	57) 8
58) 11	58) 16	58) 9	58) 6	58) 17	58) 19	58) 8	58) 6	58) 4	58) 8
59) 9	59) 20	59) 1	59) 18	59) 17	59) 11	59) 9	59) 8	59) 3	59) 7
60) 18	60) 17	60) 13	60) 6	60) 12	60) 16	60) 10	60) 6	60) 6	60) 5

ANSWER KEY

Day 11	Day 12	Day 13	Day 14	Day 15	Day 16	Day 17	Day 18	Day 19	Day 20
1) 10	1) 15	1) 11	1) 20	1) 16	1) 17	1) 10	1) 18	1) 20	1) 20
2) 5	2) 15	2) 16	2) 16	2) 16	2) 20	2) 18	2) 20	2) 8	2) 11
3) 5	3) 18	3) 19	3) 18	3) 15	3) 20	3) 10	3) 18	3) 10	3) 12
4) 10	4) 17	4) 18	4) 16	4) 19	4) 12	4) 13	4) 10	4) 12	4) 20
5) 6	5) 10	5) 16	5) 13	5) 10	5) 18	5) 14	5) 7	5) 13	5) 15
6) 10	6) 18	6) 18	6) 17	6) 9	6) 5	6) 9	6) 20	6) 8	6) 14
7) 8	7) 19	7) 3	7) 9	7) 13	7) 6	7) 4	7) 15	7) 15	7) 9
8) 8	8) 20	8) 20	8) 14	8) 4	8) 20	8) 8	8) 14	8) 12	8) 10
9) 5	9) 16	9) 17	9) 8	9) 2	9) 18	9) 17	9) 8	9) 13	9) 18
10) 10	10) 11	10) 7	10) 13	10) 19	10) 14	10) 15	10) 18	10) 6	10) 17
11) 4	11) 19	11) 14	11) 11	11) 17	11) 10	11) 11	11) 6	11) 12	11) 7
12) 6	12) 7	12) 17	12) 10	12) 8	12) 14	12) 5	12) 16	12) 16	12) 20
13) 8	13) 12	13) 19	13) 12	13) 16	13) 7	13) 7	13) 18	13) 16	13) 19
14) 4	14) 12	14) 15	14) 6	14) 20	14) 17	14) 11	14) 17	14) 14	14) 17
15) 3	15) 17	15) 8	15) 14	15) 15	15) 9	15) 5	15) 13	15) 1	15) 13
16) 9	16) 18	16) 17	16) 12	16) 20	16) 11	16) 20	16) 9	16) 19	16) 5
17) 3	17) 10	17) 14	17) 6	17) 16	17) 8	17) 10	17) 18	17) 19	17) 11
18) 9	18) 3	18) 20	18) 9	18) 19	18) 20	18) 19	18) 17	18) 8	18) 12
19) 3	19) 16	19) 18	19) 3	19) 17	19) 19	19) 3	19) 9	19) 20	19) 14
20) 9	20) 13	20) 10	20) 5	20) 11	20) 20	20) 11	20) 20	20) 10	20) 14
21) 6	21) 8	21) 8	21) 7	21) 5	21) 8	21) 19	21) 13	21) 16	21) 20
22) 1	22) 6	22) 15	22) 20	22) 14	22) 17	22) 18	22) 19	22) 13	22) 14
23) 5	23) 13	23) 18	23) 15	23) 9	23) 8	23) 14	23) 1	23) 15	23) 15
24) 5	24) 20	24) 11	24) 10	24) 20	24) 14	24) 11	24) 16	24) 9	24) 2
25) 9	25) 9	25) 5	25) 14	25) 3	25) 12	25) 15	25) 8	25) 18	25) 16
26) 8	26) 10	26) 4	26) 2	26) 2	26) 17	26) 14	26) 11	26) 18	26) 11
27) 9	27) 17	27) 14	27) 19	27) 20	27) 6	27) 20	27) 12	27) 6	27) 15
28) 10	28) 14	28) 10	28) 11	28) 17	28) 16	28) 13	28) 15	28) 11	28) 13
29) 4	29) 15	29) 11	29) 16	29) 20	29) 15	29) 11	29) 13	29) 19	29) 17
30) 7	30) 18	30) 15	30) 9	30) 20	30) 9	30) 10	30) 17	30) 4	30) 9
31) 7	31) 16	31) 4	31) 13	31) 1	31) 15	31) 19	31) 17	31) 16	31) 18
32) 8	32) 16	32) 13	32) 9	32) 13	32) 17	32) 13	32) 13	32) 17	32) 18
33) 8	33) 19	33) 8	33) 18	33) 4	33) 18	33) 6	33) 12	33) 18	33) 15
34) 6	34) 17	34) 5	34) 15	34) 12	34) 14	34) 13	34) 5	34) 18	34) 16
35) 2	35) 8	35) 13	35) 17	35) 12	35) 18	35) 18	35) 11	35) 18	35) 4
36) 7	36) 3	36) 2	36) 20	36) 12	36) 13	36) 16	36) 19	36) 20	36) 12
37) 7	37) 11	37) 13	37) 18	37) 12	37) 20	37) 15	37) 20	37) 11	37) 15
38) 6	38) 14	38) 19	38) 9	38) 12	38) 12	38) 14	38) 7	38) 20	38) 12
39) 6	39) 6	39) 19	39) 20	39) 18	39) 3	39) 11	39) 3	39) 19	39) 6
40) 4	40) 20	40) 15	40) 20	40) 19	40) 4	40) 20	40) 14	40) 15	40) 6
41) 1	41) 20	41) 9	41) 20	41) 15	41) 13	41) 17	41) 14	41) 19	41) 17
42) 0	42) 13	42) 9	42) 13	42) 19	42) 16	42) 17	42) 14	42) 3	42) 16
43) 7	43) 14	43) 7	43) 5	43) 7	43) 18	43) 15	43) 18	43) 13	43) 18
44) 9	44) 17	44) 7	44) 19	44) 16	44) 16	44) 13	44) 17	44) 3	44) 0
45) 9	45) 19	45) 9	45) 14	45) 7	45) 11	45) 20	45) 5	45) 14	45) 18
46) 4	46) 11	46) 19	46) 19	46) 4	46) 9	46) 18	46) 16	46) 9	46) 11
47) 2	47) 12	47) 19	47) 4	47) 19	47) 13	47) 3	47) 15	47) 18	47) 4
48) 9	48) 13	48) 6	48) 12	48) 12	48) 19	48) 16	48) 19	48) 20	48) 10
49) 2	49) 14	49) 19	49) 19	49) 19	49) 20	49) 13	49) 9	49) 20	49) 2
50) 8	50) 20	50) 14	50) 6	50) 16	50) 0	50) 14	50) 20	50) 8	50) 12
51) 7	51) 19	51) 5	51) 2	51) 13	51) 14	51) 19	51) 17	51) 20	51) 2
52) 10	52) 20	52) 4	52) 1	52) 12	52) 18	52) 15	52) 15	52) 20	52) 19
53) 10	53) 18	53) 8	53) 20	53) 18	53) 9	53) 9	53) 9	53) 14	53) 6
54) 7	54) 13	54) 18	54) 20	54) 10	54) 7	54) 17	54) 5	54) 13	54) 15
55) 9	55) 20	55) 11	55) 4	55) 10	55) 19	55) 5	55) 14	55) 20	55) 11
56) 7	56) 19	56) 9	56) 20	56) 17	56) 14	56) 8	56) 9	56) 15	56) 16
57) 3	57) 18	57) 15	57) 18	57) 10	57) 20	57) 19	57) 17	57) 13	57) 14
58) 10	58) 12	58) 17	58) 17	58) 19	58) 15	58) 10	58) 10	58) 12	58) 3
59) 6	59) 16	59) 10	59) 17	59) 19	59) 16	59) 14	59) 6	59) 18	59) 7
60) 8	60) 11	60) 18	60) 19	60) 6	60) 18	60) 20	60) 10	60) 7	60) 4

ANSWER KEY

Day 21	Day 22	Day 23	Day 24	Day 25	Day 26	Day 27	Day 28	Day 29	Day 30
1) 8	1) 8	1) 13	1) 19	1) 13	1) 17	1) 10	1) 8	1) 11	1) 19
2) 20	2) 20	2) 9	2) 16	2) 10	2) 15	2) 9	2) 18	2) 12	2) 20
3) 18	3) 19	3) 17	3) 19	3) 18	3) 2	3) 10	3) 20	3) 7	3) 6
4) 5	4) 10	4) 20	4) 11	4) 16	4) 15	4) 17	4) 18	4) 18	4) 8
5) 8	5) 14	5) 13	5) 10	5) 20	5) 16	5) 10	5) 19	5) 18	5) 20
6) 0	6) 9	6) 18	6) 8	6) 20	6) 12	6) 19	6) 11	6) 12	6) 3
7) 11	7) 14	7) 13	7) 19	7) 14	7) 16	7) 14	7) 18	7) 9	7) 11
8) 15	8) 16	8) 20	8) 18	8) 16	8) 15	8) 19	8) 15	8) 15	8) 18
9) 10	9) 3	9) 20	9) 14	9) 15	9) 13	9) 20	9) 4	9) 10	9) 10
10) 14	10) 20	10) 5	10) 6	10) 17	10) 11	10) 12	10) 2	10) 9	10) 18
11) 9	11) 16	11) 11	11) 7	11) 10	11) 20	11) 19	11) 20	11) 9	11) 13
12) 9	12) 5	12) 19	12) 17	12) 14	12) 19	12) 12	12) 15	12) 14	12) 19
13) 9	13) 14	13) 17	13) 16	13) 8	13) 19	13) 13	13) 8	13) 16	13) 13
14) 15	14) 20	14) 13	14) 4	14) 9	14) 4	14) 11	14) 12	14) 20	14) 13
15) 13	15) 16	15) 11	15) 17	15) 17	15) 17	15) 13	15) 4	15) 11	15) 9
16) 20	16) 10	16) 18	16) 14	16) 16	16) 5	16) 19	16) 15	16) 4	16) 14
17) 13	17) 20	17) 17	17) 13	17) 20	17) 20	17) 20	17) 17	17) 13	17) 20
18) 20	18) 16	18) 10	18) 10	18) 2	18) 11	18) 15	18) 14	18) 13	18) 17
19) 14	19) 16	19) 6	19) 3	19) 17	19) 20	19) 8	19) 20	19) 11	19) 17
20) 13	20) 20	20) 12	20) 12	20) 14	20) 20	20) 17	20) 17	20) 18	20) 3
21) 15	21) 3	21) 8	21) 13	21) 16	21) 15	21) 14	21) 19	21) 2	21) 11
22) 11	22) 16	22) 17	22) 7	22) 6	22) 16	22) 20	22) 7	22) 13	22) 13
23) 14	23) 3	23) 8	23) 18	23) 4	23) 19	23) 10	23) 19	23) 14	23) 16
24) 18	24) 11	24) 20	24) 6	24) 15	24) 12	24) 14	24) 18	24) 10	24) 8
25) 13	25) 15	25) 13	25) 13	25) 10	25) 18	25) 9	25) 18	25) 20	25) 4
26) 18	26) 19	26) 18	26) 12	26) 12	26) 7	26) 7	26) 16	26) 18	26) 10
27) 12	27) 7	27) 19	27) 20	27) 14	27) 18	27) 14	27) 10	27) 6	27) 10
28) 15	28) 16	28) 16	28) 13	28) 18	28) 19	28) 7	28) 15	28) 16	28) 8
29) 4	29) 14	29) 12	29) 17	29) 13	29) 1	29) 19	29) 20	29) 17	29) 12
30) 11	30) 6	30) 11	30) 18	30) 8	30) 18	30) 20	30) 19	30) 8	30) 14
31) 15	31) 19	31) 8	31) 14	31) 19	31) 20	31) 13	31) 16	31) 11	31) 3
32) 6	32) 2	32) 19	32) 15	32) 12	32) 10	32) 12	32) 20	32) 17	32) 11
33) 2	33) 9	33) 6	33) 9	33) 18	33) 16	33) 9	33) 16	33) 19	33) 13
34) 5	34) 5	34) 10	34) 18	34) 19	34) 20	34) 17	34) 8	34) 12	34) 20
35) 18	35) 9	35) 7	35) 9	35) 18	35) 11	35) 8	35) 10	35) 13	35) 4
36) 15	36) 19	36) 16	36) 18	36) 5	36) 12	36) 12	36) 15	36) 14	36) 7
37) 18	37) 14	37) 14	37) 16	37) 0	37) 12	37) 17	37) 15	37) 9	37) 0
38) 14	38) 15	38) 19	38) 16	38) 8	38) 17	38) 1	38) 4	38) 6	38) 12
39) 12	39) 17	39) 16	39) 18	39) 5	39) 16	39) 20	39) 17	39) 2	39) 19
40) 19	40) 3	40) 17	40) 20	40) 14	40) 15	40) 17	40) 18	40) 18	40) 18
41) 17	41) 8	41) 17	41) 9	41) 17	41) 11	41) 13	41) 1	41) 7	41) 16
42) 18	42) 13	42) 2	42) 19	42) 12	42) 19	42) 14	42) 19	42) 15	42) 6
43) 8	43) 20	43) 13	43) 16	43) 3	43) 15	43) 14	43) 20	43) 16	43) 16
44) 18	44) 4	44) 17	44) 6	44) 20	44) 10	44) 16	44) 17	44) 16	44) 20
45) 9	45) 12	45) 16	45) 18	45) 13	45) 20	45) 5	45) 6	45) 16	45) 11
46) 14	46) 12	46) 7	46) 10	46) 8	46) 19	46) 15	46) 12	46) 6	46) 20
47) 10	47) 18	47) 12	47) 8	47) 20	47) 10	47) 8	47) 18	47) 19	47) 15
48) 20	48) 6	48) 19	48) 19	48) 13	48) 15	48) 9	48) 7	48) 10	48) 14
49) 10	49) 4	49) 16	49) 1	49) 15	49) 19	49) 15	49) 9	49) 16	49) 6
50) 4	50) 14	50) 17	50) 8	50) 17	50) 7	50) 14	50) 6	50) 14	50) 10
51) 17	51) 17	51) 7	51) 19	51) 19	51) 13	51) 13	51) 14	51) 18	51) 14
52) 10	52) 12	52) 13	52) 13	52) 14	52) 16	52) 16	52) 19	52) 11	52) 18
53) 17	53) 12	53) 16	53) 11	53) 9	53) 9	53) 16	53) 20	53) 15	53) 15
54) 17	54) 19	54) 15	54) 17	54) 5	54) 17	54) 18	54) 17	54) 3	54) 17
55) 10	55) 17	55) 16	55) 14	55) 18	55) 11	55) 20	55) 20	55) 5	55) 4
56) 20	56) 12	56) 15	56) 11	56) 10	56) 8	56) 8	56) 18	56) 16	56) 18
57) 19	57) 19	57) 18	57) 20	57) 4	57) 16	57) 5	57) 6	57) 17	57) 17
58) 16	58) 15	58) 16	58) 7	58) 20	58) 17	58) 19	58) 5	58) 19	58) 18
59) 7	59) 14	59) 15	59) 19	59) 11	59) 15	59) 7	59) 11	59) 12	59) 3
60) 13	60) 14	60) 18	60) 14	60) 18	60) 6	60) 19	60) 10	60) 14	60) 13

ANSWER KEY

Day 31	Day 32	Day 33	Day 34	Day 35	Day 36	Day 37	Day 38	Day 39	Day 40
1) 1	1) 20	1) 14	1) 2	1) 9	1) 19	1) 6	1) 0	1) 0	1) 0
2) 12	2) 20	2) 10	2) 13	2) 19	2) 1	2) 3	2) 2	2) 2	2) 0
3) 17	3) 19	3) 17	3) 11	3) 1	3) 2	3) 1	3) 6	3) 5	3) 3
4) 15	4) 20	4) 7	4) 5	4) 5	4) 10	4) 9	4) 9	4) 4	4) 3
5) 9	5) 14	5) 0	5) 16	5) 1	5) 3	5) 7	5) 8	5) 4	5) 3
6) 16	6) 9	6) 5	6) 9	6) 4	6) 16	6) 8	6) 7	6) 9	6) 4
7) 20	7) 8	7) 6	7) 0	7) 6	7) 6	7) 4	7) 10	7) 3	7) 8
8) 19	8) 15	8) 13	8) 17	8) 14	8) 11	8) 5	8) 9	8) 5	8) 5
9) 18	9) 12	9) 20	9) 3	9) 17	9) 13	9) 10	9) 3	9) 0	9) 7
10) 13	10) 16	10) 4	10) 12	10) 18	10) 18	10) 9	10) 7	10) 5	10) 8
11) 7	11) 16	11) 8	11) 19	11) 12	11) 15	11) 10	11) 2	11) 3	11) 4
12) 15	12) 13	12) 12	12) 1	12) 3	12) 8	12) 5	12) 5	12) 3	12) 1
13) 10	13) 17	13) 18	13) 6	13) 13	13) 4	13) 3	13) 1	13) 5	13) 2
14) 15	14) 1	14) 11	14) 7	14) 16	14) 14	14) 1	14) 1	14) 4	14) 7
15) 18	15) 19	15) 19	15) 14	15) 11	15) 9	15) 0	15) 5	15) 1	15) 4
16) 17	16) 8	16) 9	16) 8	16) 15	16) 7	16) 7	16) 6	16) 3	16) 1
17) 13	17) 18	17) 16	17) 10	17) 0	17) 0	17) 2	17) 4	17) 1	17) 0
18) 8	18) 19	18) 1	18) 4	18) 8	18) 17	18) 4	18) 10	18) 6	18) 0
19) 12	19) 12	19) 15	19) 20	19) 10	19) 12	19) 8	19) 3	19) 0	19) 8
20) 19	20) 15	20) 2	20) 18	20) 2	20) 5	20) 6	20) 8	20) 10	20) 4
21) 15	21) 10	21) 3	21) 15	21) 7	21) 1	21) 2	21) 4	21) 4	21) 0
22) 17	22) 11	22) 11	22) 8	22) 11	22) 8	22) 4	22) 5	22) 1	22) 6
23) 7	23) 9	23) 7	23) 13	23) 15	23) 11	23) 9	23) 6	23) 9	23) 1
24) 11	24) 20	24) 0	24) 16	24) 8	24) 17	24) 2	24) 10	24) 8	24) 2
25) 13	25) 4	25) 18	25) 4	25) 3	25) 4	25) 7	25) 3	25) 4	25) 5
26) 17	26) 19	26) 14	26) 19	26) 1	26) 1	26) 1	26) 9	26) 2	26) 3
27) 11	27) 12	27) 6	27) 14	27) 14	27) 9	27) 8	27) 9	27) 2	27) 3
28) 5	28) 13	28) 16	28) 10	28) 13	28) 14	28) 3	28) 8	28) 1	28) 1
29) 18	29) 17	29) 10	29) 7	29) 1	29) 19	29) 5	29) 2	29) 7	29) 6
30) 17	30) 14	30) 3	30) 15	30) 12	30) 0	30) 2	30) 1	30) 6	30) 2
31) 9	31) 9	31) 13	31) 1	31) 10	31) 7	31) 7	31) 7	31) 0	31) 9
32) 17	32) 19	32) 1	32) 12	32) 7	32) 10	32) 6	32) 8	32) 2	32) 0
33) 5	33) 20	33) 9	33) 5	33) 17	33) 1	33) 6	33) 3	33) 5	33) 4
34) 17	34) 18	34) 17	34) 6	34) 5	34) 3	34) 10	34) 7	34) 7	34) 6
35) 18	35) 8	35) 4	35) 2	35) 0	35) 16	35) 1	35) 10	35) 0	35) 3
36) 14	36) 20	36) 19	36) 3	36) 16	36) 5	36) 5	36) 5	36) 2	36) 2
37) 9	37) 20	37) 8	37) 11	37) 19	37) 2	37) 3	37) 6	37) 4	37) 5
38) 19	38) 19	38) 15	38) 0	38) 18	38) 18	38) 0	38) 0	38) 2	38) 0
39) 19	39) 2	39) 20	39) 20	39) 9	39) 12	39) 10	39) 4	39) 7	39) 5
40) 18	40) 20	40) 5	40) 17	40) 4	40) 13	40) 9	40) 1	40) 6	40) 1
41) 2	41) 4	41) 2	41) 18	41) 2	41) 15	41) 8	41) 4	41) 1	41) 1
42) 14	42) 15	42) 12	42) 9	42) 6	42) 6	42) 4	42) 2	42) 2	42) 3
43) 10	43) 11	43) 1	43) 15	43) 8	43) 8	43) 9	43) 4	43) 1	43) 2
44) 18	44) 7	44) 10	44) 13	44) 1	44) 3	44) 5	44) 7	44) 8	44) 3
45) 14	45) 8	45) 20	45) 5	45) 15	45) 15	45) 4	45) 8	45) 6	45) 5
46) 6	46) 11	46) 18	46) 0	46) 7	46) 6	46) 10	46) 10	46) 0	46) 2
47) 17	47) 14	47) 4	47) 18	47) 4	47) 16	47) 7	47) 3	47) 3	47) 2
48) 10	48) 18	48) 19	48) 8	48) 13	48) 14	48) 6	48) 2	48) 1	48) 4
49) 16	49) 11	49) 7	49) 7	49) 10	49) 19	49) 9	49) 1	49) 8	49) 4
50) 13	50) 16	50) 0	50) 1	50) 17	50) 5	50) 0	50) 7	50) 4	50) 0
51) 17	51) 16	51) 6	51) 14	51) 3	51) 1	51) 8	51) 8	51) 0	51) 5
52) 16	52) 20	52) 3	52) 16	52) 6	52) 0	52) 7	52) 5	52) 0	52) 1
53) 12	53) 16	53) 5	53) 17	53) 14	53) 1	53) 4	53) 5	53) 6	53) 0
54) 18	54) 16	54) 14	54) 19	54) 12	54) 7	54) 2	54) 6	54) 7	54) 0
55) 12	55) 16	55) 17	55) 11	55) 9	55) 10	55) 1	55) 1	55) 3	55) 6
56) 10	56) 15	56) 9	56) 3	56) 11	56) 2	56) 3	56) 2	56) 3	56) 7
57) 8	57) 18	57) 2	57) 20	57) 16	57) 9	57) 5	57) 3	57) 1	57) 0
58) 15	58) 12	58) 16	58) 12	58) 1	58) 17	58) 8	58) 10	58) 1	58) 2
59) 9	59) 6	59) 8	59) 6	59) 18	59) 13	59) 6	59) 4	59) 2	59) 10
60) 19	60) 5	60) 12	60) 9	60) 2	60) 18	60) 2	60) 9	60) 0	60) 9

ANSWER KEY

	Day 41	Day 42	Day 43	Day 44	Day 45	Day 46	Day 47	Day 48	Day 49	Day 50
1)	0	7	2	3	0	3	9	1	5	2
2)	6	7	7	1	2	7	0	4	1	2
3)	3	7	2	7	4	5	0	8	3	5
4)	0	3	2	7	3	11	6	14	2	18
5)	8	1	1	14	10	2	3	5	6	4
6)	2	5	2	3	10	11	1	17	11	17
7)	7	1	4	4	13	6	12	9	4	8
8)	2	1	3	15	6	9	14	2	10	17
9)	5	2	6	9	10	13	1	19	17	4
10)	5	4	4	6	8	2	11	15	3	0
11)	4	7	1	5	5	2	6	0	7	13
12)	10	6	0	1	4	6	4	7	2	5
13)	1	4	0	16	1	1	2	9	11	12
14)	0	2	5	20	17	3	5	5	1	0
15)	3	2	2	2	2	16	3	7	3	2
16)	2	1	3	1	15	2	12	10	4	16
17)	1	0	3	10	7	8	7	13	6	0
18)	7	2	1	11	3	12	2	8	8	8
19)	7	1	0	0	6	5	3	3	9	7
20)	1	3	2	0	9	6	1	1	5	4
21)	1	1	4	12	7	3	1	4	3	7
22)	0	8	2	11	1	14	6	4	13	15
23)	4	0	2	7	17	8	10	8	3	13
24)	1	3	0	3	1	3	6	12	10	10
25)	1	2	0	18	3	17	0	2	4	14
26)	6	1	5	0	5	3	0	1	7	0
27)	2	3	3	13	1	4	2	8	8	5
28)	2	5	3	10	6	7	2	11	10	9
29)	6	4	1	2	7	3	4	1	6	14
30)	7	3	8	17	15	6	10	18	0	0
31)	1	9	7	9	11	7	2	9	9	11
32)	9	10	0	8	1	9	12	13	6	7
33)	9	0	3	12	9	0	7	11	0	5
34)	0	4	4	9	6	3	4	2	0	14
35)	2	6	3	4	8	0	19	12	6	8
36)	5	0	1	16	12	15	16	2	9	8
37)	4	5	6	1	10	4	7	8	7	3
38)	0	1	6	4	5	5	13	1	8	0
39)	1	0	7	0	7	2	3	3	4	0
40)	3	4	0	8	2	13	6	11	14	8
41)	0	0	10	12	1	5	8	6	1	1
42)	3	1	9	14	16	4	6	8	14	3
43)	5	9	4	8	11	5	1	1	11	9
44)	4	4	7	9	3	4	15	16	4	5
45)	0	6	1	1	4	8	5	3	0	4
46)	0	2	1	18	0	16	16	5	5	3
47)	3	3	5	10	2	11	12	2	11	11
48)	3	5	1	4	0	4	12	15	0	3
49)	5	1	0	10	14	12	11	3	6	3
50)	8	2	6	9	0	2	17	8	4	0
51)	2	0	1	3	13	0	13	8	4	3
52)	3	0	5	5	0	7	3	11	4	7
53)	4	4	6	6	3	2	4	4	12	9
54)	8	5	2	2	0	15	5	8	10	5
55)	1	2	4	2	4	5	7	0	0	4
56)	0	0	1	10	1	19	2	14	17	12
57)	2	3	5	1	2	9	2	0	14	15
58)	4	6	0	14	15	4	9	2	14	8
59)	2	5	0	6	7	8	9	6	5	2
60)	0	6	0	9	0	0	5	7	15	17

ANSWER KEY

KEY

#	Day 51	Day 52	Day 53	Day 54	Day 55	Day 56	Day 57	Day 58	Day 59	Day 60
1)	8	6	3	10	20	2	2	3	2	1
2)	0	12	1	0	1	14	5	6	3	6
3)	18	3	7	0	15	8	12	3	9	8
4)	2	12	12	6	10	2	7	17	15	3
5)	9	14	1	9	5	6	11	6	5	6
6)	3	19	5	4	8	3	10	16	3	2
7)	13	2	9	6	2	4	7	12	0	11
8)	0	8	3	5	8	9	5	7	8	12
9)	6	1	9	18	7	0	4	18	0	8
10)	2	0	7	12	1	3	1	7	5	14
11)	0	4	1	3	14	4	4	9	12	3
12)	3	0	4	6	1	2	13	7	5	8
13)	3	5	8	11	11	13	1	4	11	0
14)	10	4	2	4	8	6	0	15	2	2
15)	12	7	6	7	17	9	2	0	2	4
16)	8	9	7	7	2	5	0	3	5	8
17)	10	6	11	0	9	13	4	0	4	9
18)	6	10	0	5	6	8	8	0	16	1
19)	2	3	4	3	2	2	11	0	11	1
20)	6	12	0	7	10	9	19	0	3	4
21)	12	1	19	0	5	15	4	0	13	10
22)	2	14	6	3	17	7	14	16	16	8
23)	9	1	7	7	16	10	5	5	10	3
24)	1	11	3	2	4	4	17	3	2	11
25)	0	5	5	3	19	8	5	0	11	1
26)	7	10	1	12	15	0	3	4	18	12
27)	0	8	16	7	5	10	8	10	10	0
28)	8	14	0	9	9	2	6	12	9	16
29)	6	2	11	1	14	3	5	7	1	13
30)	1	1	10	1	0	17	1	2	2	14
31)	11	0	14	8	13	12	6	10	0	0
32)	7	16	10	1	4	6	0	11	7	4
33)	10	2	1	6	12	6	13	8	6	0
34)	7	2	16	3	4	9	12	3	14	2
35)	10	2	3	3	9	11	6	10	5	14
36)	1	9	4	7	1	8	9	1	10	13
37)	9	2	1	2	2	15	1	9	10	3
38)	0	2	8	6	6	4	7	1	6	1
39)	6	13	13	1	2	14	8	14	7	4
40)	3	5	14	11	2	1	6	14	0	17
41)	15	18	2	4	1	9	7	2	0	6
42)	7	5	5	0	10	2	3	10	1	15
43)	4	13	1	10	13	0	6	1	9	2
44)	6	7	2	11	0	1	7	0	2	9
45)	16	5	13	8	0	1	0	12	2	2
46)	1	4	14	15	6	5	6	15	4	17
47)	20	8	0	0	2	18	10	1	3	14
48)	11	6	4	5	0	3	5	10	1	4
49)	9	4	3	10	0	1	10	2	9	6
50)	10	2	12	0	3	5	9	2	1	2
51)	13	1	1	14	15	14	2	1	7	4
52)	1	19	10	0	8	0	4	3	13	1
53)	11	14	0	2	5	13	5	15	7	11
54)	11	2	9	4	16	9	3	13	5	2
55)	3	9	0	12	11	5	16	0	5	6
56)	5	4	9	0	11	11	9	7	3	0
57)	6	4	4	14	5	0	13	4	11	4
58)	1	0	5	6	1	4	12	11	0	4
59)	8	5	13	2	8	20	12	11	12	6
60)	6	16	12	13	1	7	15	0	16	13

ANSWER KEY

KEY

	Day 61	Day 62	Day 63	Day 64	Day 65	Day 66	Day 67	Day 68	Day 69	Day 70
1)	1	2	17	4	16	18	0	16	4	20
2)	7	10	5	11	2	4	18	12	2	8
3)	11	13	5	8	2	16	16	6	2	12
4)	20	1	8	8	6	12	10	14	6	10
5)	5	0	0	3	2	6	8	8	4	8
6)	0	7	5	4	10	6	10	12	12	14
7)	10	1	7	4	12	0	6	16	4	16
8)	4	10	11	4	16	18	2	2	12	6
9)	1	8	10	0	6	2	18	0	18	0
10)	14	3	0	6	16	18	0	4	16	2
11)	13	2	5	3	14	2	20	8	16	0
12)	0	9	7	19	8	14	0	4	12	6
13)	14	3	4	12	20	20	6	0	14	12
14)	7	5	18	8	20	2	14	0	0	4
15)	12	5	1	12	0	0	18	8	6	2
16)	5	15	20	0	4	12	12	8	6	10
17)	6	16	14	7	2	20	8	18	2	4
18)	8	9	2	1	18	0	0	12	20	0
19)	7	18	4	2	16	8	0	14	2	2
20)	19	4	7	9	12	16	8	20	0	4
21)	2	4	2	15	10	6	12	12	8	18
22)	13	10	7	5	8	14	10	8	16	18
23)	3	17	6	12	8	20	6	10	20	16
24)	18	5	8	8	14	14	10	18	12	14
25)	6	14	4	2	6	14	6	0	12	12
26)	19	6	1	4	4	12	8	18	10	4
27)	8	1	12	12	4	14	18	2	14	10
28)	3	3	4	6	16	14	16	14	2	6
29)	1	9	7	1	20	18	8	14	14	16
30)	8	3	2	13	18	10	18	4	10	16
31)	5	1	3	0	12	16	0	20	0	6
32)	3	7	15	1	12	6	0	10	4	18
33)	6	19	9	0	2	4	10	6	16	10
34)	4	11	10	5	4	8	8	10	14	14
35)	9	8	9	11	20	18	4	20	10	6
36)	10	2	9	16	20	8	16	6	20	16
37)	17	3	5	2	16	12	10	0	6	10
38)	1	5	10	0	12	20	4	8	8	8
39)	2	1	0	15	8	14	2	10	14	8
40)	4	14	2	1	2	4	2	2	8	14
41)	9	5	8	0	10	10	14	18	20	0
42)	1	7	9	11	8	4	4	2	2	10
43)	15	1	12	2	20	0	10	4	18	2
44)	6	9	1	15	6	16	12	10	2	8
45)	3	1	13	14	0	20	12	8	2	4
46)	9	3	7	5	10	12	8	20	6	14
47)	2	10	3	7	8	10	20	6	10	4
48)	0	12	6	3	20	12	14	6	2	4
49)	7	6	2	0	10	10	16	6	18	12
50)	7	11	6	1	16	6	18	12	16	14
51)	4	8	0	2	14	4	6	16	14	6
52)	17	18	4	2	20	10	14	18	8	12
53)	1	0	4	3	20	14	18	16	20	20
54)	8	13	11	7	12	6	2	2	0	0
55)	10	9	0	4	0	2	20	14	2	14
56)	11	2	14	6	4	12	4	16	0	10
57)	2	2	6	9	14	10	8	20	0	12
58)	3	15	0	11	18	2	14	18	16	20
59)	9	8	2	0	6	10	10	14	20	20
60)	0	4	2	3	8	6	12	18	6	20

ANSWER KEY

Day 71	Day 72	Day 73	Day 74	Day 75	Day 76	Day 77	Day 78	Day 79	Day 80
1) 4	1) 4	1) 12	1) 4	1) 4	1) 13	1) 20	1) 9	1) 1	1) 6
2) 14	2) 0	2) 10	2) 20	2) 17	2) 7	2) 16	2) 4	2) 7	2) 15
3) 12	3) 14	3) 14	3) 14	3) 5	3) 7	3) 1	3) 1	3) 5	3) 9
4) 6	4) 8	4) 16	4) 14	4) 10	4) 4	4) 16	4) 16	4) 8	4) 13
5) 10	5) 6	5) 2	5) 4	5) 16	5) 11	5) 10	5) 11	5) 3	5) 10
6) 6	6) 2	6) 2	6) 8	6) 16	6) 11	6) 14	6) 17	6) 6	6) 14
7) 16	7) 4	7) 14	7) 16	7) 15	7) 20	7) 19	7) 7	7) 20	7) 2
8) 2	8) 20	8) 10	8) 8	8) 18	8) 14	8) 4	8) 18	8) 13	8) 9
9) 14	9) 8	9) 12	9) 18	9) 16	9) 2	9) 8	9) 13	9) 1	9) 12
10) 16	10) 18	10) 6	10) 18	10) 13	10) 6	10) 19	10) 5	10) 19	10) 17
11) 0	11) 8	11) 20	11) 4	11) 15	11) 0	11) 4	11) 15	11) 1	11) 10
12) 20	12) 4	12) 20	12) 2	12) 3	12) 12	12) 6	12) 11	12) 5	12) 15
13) 2	13) 12	13) 16	13) 18	13) 1	13) 14	13) 6	13) 8	13) 13	13) 19
14) 20	14) 14	14) 10	14) 20	14) 9	14) 1	14) 10	14) 18	14) 17	14) 0
15) 20	15) 18	15) 4	15) 10	15) 20	15) 6	15) 17	15) 18	15) 12	15) 14
16) 10	16) 4	16) 12	16) 4	16) 4	16) 19	16) 6	16) 9	16) 14	16) 15
17) 10	17) 6	17) 2	17) 14	17) 5	17) 6	17) 5	17) 6	17) 5	17) 5
18) 2	18) 12	18) 0	18) 12	18) 5	18) 13	18) 16	18) 15	18) 5	18) 18
19) 12	19) 18	19) 18	19) 18	19) 20	19) 2	19) 15	19) 17	19) 12	19) 11
20) 12	20) 20	20) 2	20) 16	20) 11	20) 6	20) 9	20) 20	20) 8	20) 13
21) 14	21) 16	21) 4	21) 14	21) 0	21) 18	21) 18	21) 15	21) 15	21) 1
22) 6	22) 6	22) 14	22) 8	22) 11	22) 20	22) 19	22) 14	22) 12	22) 7
23) 18	23) 10	23) 20	23) 0	23) 7	23) 18	23) 11	23) 5	23) 0	23) 18
24) 4	24) 18	24) 10	24) 10	24) 19	24) 16	24) 10	24) 4	24) 10	24) 7
25) 18	25) 2	25) 12	25) 10	25) 17	25) 3	25) 20	25) 9	25) 8	25) 17
26) 16	26) 4	26) 6	26) 18	26) 11	26) 14	26) 15	26) 18	26) 20	26) 19
27) 6	27) 20	27) 18	27) 16	27) 15	27) 20	27) 1	27) 12	27) 11	27) 20
28) 8	28) 10	28) 8	28) 16	28) 20	28) 12	28) 15	28) 17	28) 6	28) 18
29) 4	29) 0	29) 12	29) 18	29) 3	29) 16	29) 16	29) 18	29) 12	29) 16
30) 0	30) 8	30) 10	30) 18	30) 17	30) 3	30) 19	30) 12	30) 6	30) 1
31) 14	31) 14	31) 20	31) 14	31) 5	31) 0	31) 8	31) 15	31) 8	31) 15
32) 0	32) 10	32) 0	32) 12	32) 3	32) 2	32) 11	32) 11	32) 19	32) 10
33) 14	33) 12	33) 18	33) 6	33) 4	33) 18	33) 1	33) 16	33) 4	33) 16
34) 2	34) 6	34) 16	34) 10	34) 9	34) 9	34) 10	34) 13	34) 4	34) 11
35) 8	35) 16	35) 2	35) 10	35) 19	35) 10	35) 16	35) 5	35) 6	35) 3
36) 18	36) 2	36) 16	36) 6	36) 2	36) 10	36) 16	36) 3	36) 13	36) 2
37) 2	37) 16	37) 20	37) 4	37) 15	37) 1	37) 12	37) 9	37) 19	37) 2
38) 10	38) 4	38) 6	38) 0	38) 2	38) 4	38) 3	38) 4	38) 2	38) 14
39) 0	39) 16	39) 16	39) 10	39) 6	39) 6	39) 8	39) 20	39) 13	39) 7
40) 10	40) 0	40) 6	40) 20	40) 5	40) 8	40) 13	40) 1	40) 5	40) 6
41) 10	41) 8	41) 0	41) 6	41) 7	41) 17	41) 20	41) 4	41) 7	41) 18
42) 10	42) 16	42) 12	42) 14	42) 20	42) 12	42) 1	42) 17	42) 6	42) 0
43) 4	43) 6	43) 8	43) 12	43) 8	43) 11	43) 2	43) 13	43) 14	43) 3
44) 12	44) 20	44) 4	44) 6	44) 19	44) 16	44) 8	44) 8	44) 10	44) 5
45) 2	45) 2	45) 4	45) 8	45) 6	45) 1	45) 0	45) 1	45) 17	45) 5
46) 16	46) 10	46) 20	46) 20	46) 9	46) 7	46) 4	46) 11	46) 20	46) 14
47) 0	47) 6	47) 2	47) 0	47) 12	47) 11	47) 12	47) 12	47) 2	47) 4
48) 6	48) 14	48) 2	48) 4	48) 14	48) 1	48) 9	48) 0	48) 7	48) 2
49) 8	49) 18	49) 16	49) 2	49) 0	49) 1	49) 4	49) 10	49) 12	49) 2
50) 4	50) 8	50) 16	50) 0	50) 17	50) 7	50) 12	50) 20	50) 15	50) 18
51) 4	51) 0	51) 0	51) 20	51) 17	51) 19	51) 7	51) 8	51) 13	51) 17
52) 8	52) 10	52) 16	52) 6	52) 3	52) 19	52) 18	52) 2	52) 9	52) 20
53) 8	53) 14	53) 8	53) 8	53) 3	53) 1	53) 15	53) 14	53) 20	53) 8
54) 4	54) 8	54) 6	54) 4	54) 16	54) 2	54) 7	54) 6	54) 2	54) 7
55) 6	55) 8	55) 14	55) 0	55) 7	55) 19	55) 15	55) 19	55) 14	55) 14
56) 0	56) 14	56) 14	56) 0	56) 13	56) 4	56) 9	56) 0	56) 4	56) 13
57) 18	57) 18	57) 8	57) 10	57) 5	57) 7	57) 17	57) 0	57) 5	57) 9
58) 20	58) 6	58) 4	58) 8	58) 3	58) 17	58) 14	58) 4	58) 11	58) 9
59) 14	59) 20	59) 6	59) 10	59) 14	59) 0	59) 3	59) 19	59) 11	59) 15
60) 16	60) 14	60) 6	60) 20	60) 12	60) 4	60) 10	60) 17	60) 20	60) 11

ANSWER KEY

KEY

Day 81	Day 82	Day 83	Day 84	Day 85	Day 86	Day 87	Day 88	Day 89	Day 90
1) 18	1) 10	1) 11	1) 17	1) 3	1) 6	1) 17	1) 19	1) 16	1) 8
2) 13	2) 10	2) 19	2) 9	2) 12	2) 16	2) 19	2) 2	2) 0	2) 11
3) 1	3) 7	3) 20	3) 6	3) 15	3) 14	3) 4	3) 15	3) 11	3) 14
4) 13	4) 6	4) 14	4) 4	4) 2	4) 5	4) 8	4) 13	4) 5	4) 6
5) 4	5) 9	5) 15	5) 1	5) 8	5) 4	5) 6	5) 16	5) 11	5) 19
6) 8	6) 19	6) 12	6) 10	6) 18	6) 16	6) 13	6) 0	6) 13	6) 1
7) 14	7) 7	7) 11	7) 13	7) 10	7) 6	7) 19	7) 8	7) 7	7) 16
8) 8	8) 17	8) 13	8) 11	8) 5	8) 4	8) 1	8) 12	8) 19	8) 16
9) 19	9) 12	9) 15	9) 20	9) 4	9) 0	9) 14	9) 11	9) 15	9) 14
10) 2	10) 2	10) 6	10) 15	10) 9	10) 19	10) 13	10) 4	10) 17	10) 16
11) 14	11) 5	11) 18	11) 0	11) 3	11) 17	11) 8	11) 20	11) 2	11) 3
12) 0	12) 8	12) 1	12) 2	12) 9	12) 8	12) 7	12) 20	12) 6	12) 13
13) 12	13) 5	13) 14	13) 8	13) 19	13) 12	13) 8	13) 18	13) 16	13) 7
14) 7	14) 7	14) 15	14) 7	14) 3	14) 14	14) 3	14) 10	14) 4	14) 6
15) 3	15) 9	15) 5	15) 0	15) 15	15) 8	15) 17	15) 5	15) 19	15) 5
16) 3	16) 4	16) 13	16) 0	16) 5	16) 10	16) 3	16) 3	16) 1	16) 18
17) 3	17) 14	17) 9	17) 3	17) 17	17) 11	17) 0	17) 20	17) 10	17) 15
18) 14	18) 19	18) 4	18) 5	18) 17	18) 9	18) 8	18) 10	18) 15	18) 7
19) 10	19) 8	19) 2	19) 7	19) 11	19) 1	19) 2	19) 3	19) 15	19) 1
20) 20	20) 16	20) 20	20) 13	20) 9	20) 2	20) 16	20) 18	20) 7	20) 5
21) 17	21) 6	21) 9	21) 1	21) 6	21) 19	21) 2	21) 7	21) 0	21) 19
22) 11	22) 7	22) 9	22) 14	22) 4	22) 4	22) 4	22) 11	22) 2	22) 7
23) 12	23) 10	23) 4	23) 6	23) 12	23) 12	23) 18	23) 6	23) 5	23) 17
24) 2	24) 5	24) 0	24) 12	24) 18	24) 16	24) 16	24) 17	24) 1	24) 0
25) 0	25) 20	25) 15	25) 4	25) 13	25) 16	25) 12	25) 9	25) 20	25) 0
26) 11	26) 6	26) 15	26) 10	26) 5	26) 9	26) 15	26) 13	26) 20	26) 5
27) 6	27) 7	27) 3	27) 16	27) 2	27) 16	27) 2	27) 14	27) 3	27) 12
28) 7	28) 12	28) 14	28) 16	28) 15	28) 20	28) 19	28) 19	28) 3	28) 6
29) 9	29) 1	29) 3	29) 18	29) 6	29) 10	29) 9	29) 20	29) 1	29) 13
30) 19	30) 10	30) 10	30) 6	30) 10	30) 5	30) 17	30) 10	30) 9	30) 18
31) 12	31) 4	31) 11	31) 10	31) 16	31) 4	31) 19	31) 15	31) 19	31) 16
32) 2	32) 13	32) 13	32) 6	32) 11	32) 18	32) 18	32) 7	32) 14	32) 10
33) 17	33) 19	33) 17	33) 1	33) 6	33) 12	33) 19	33) 18	33) 15	33) 20
34) 10	34) 3	34) 2	34) 14	34) 13	34) 11	34) 5	34) 0	34) 0	34) 18
35) 14	35) 0	35) 20	35) 20	35) 2	35) 10	35) 14	35) 6	35) 6	35) 0
36) 13	36) 18	36) 14	36) 19	36) 11	36) 7	36) 4	36) 17	36) 8	36) 4
37) 6	37) 2	37) 17	37) 19	37) 8	37) 18	37) 16	37) 7	37) 14	37) 19
38) 3	38) 8	38) 9	38) 5	38) 9	38) 9	38) 12	38) 1	38) 1	38) 7
39) 9	39) 2	39) 5	39) 5	39) 13	39) 7	39) 2	39) 16	39) 13	39) 6
40) 2	40) 16	40) 0	40) 15	40) 13	40) 14	40) 1	40) 9	40) 4	40) 9
41) 3	41) 12	41) 14	41) 16	41) 11	41) 15	41) 13	41) 12	41) 12	41) 3
42) 2	42) 2	42) 14	42) 13	42) 14	42) 13	42) 5	42) 11	42) 14	42) 18
43) 8	43) 12	43) 0	43) 0	43) 5	43) 2	43) 17	43) 0	43) 7	43) 2
44) 18	44) 1	44) 14	44) 3	44) 6	44) 0	44) 18	44) 15	44) 14	44) 8
45) 1	45) 10	45) 1	45) 6	45) 2	45) 20	45) 6	45) 17	45) 15	45) 9
46) 10	46) 19	46) 17	46) 1	46) 3	46) 14	46) 9	46) 8	46) 16	46) 16
47) 3	47) 11	47) 7	47) 7	47) 7	47) 5	47) 7	47) 16	47) 16	47) 11
48) 10	48) 12	48) 11	48) 12	48) 3	48) 20	48) 20	48) 19	48) 3	48) 18
49) 20	49) 7	49) 3	49) 9	49) 12	49) 1	49) 15	49) 5	49) 20	49) 9
50) 0	50) 0	50) 4	50) 7	50) 20	50) 11	50) 17	50) 20	50) 6	50) 10
51) 2	51) 11	51) 19	51) 13	51) 10	51) 8	51) 18	51) 13	51) 4	51) 10
52) 19	52) 0	52) 18	52) 11	52) 3	52) 19	52) 13	52) 1	52) 17	52) 6
53) 17	53) 3	53) 5	53) 12	53) 20	53) 14	53) 1	53) 18	53) 5	53) 4
54) 7	54) 4	54) 3	54) 8	54) 15	54) 18	54) 8	54) 2	54) 17	54) 6
55) 0	55) 16	55) 9	55) 17	55) 17	55) 19	55) 2	55) 2	55) 14	55) 8
56) 0	56) 18	56) 20	56) 11	56) 4	56) 12	56) 7	56) 13	56) 17	56) 13
57) 0	57) 10	57) 19	57) 0	57) 16	57) 14	57) 5	57) 3	57) 3	57) 19
58) 16	58) 10	58) 0	58) 18	58) 8	58) 6	58) 0	58) 15	58) 11	58) 11
59) 7	59) 20	59) 18	59) 13	59) 6	59) 1	59) 8	59) 15	59) 0	59) 4
60) 8	60) 11	60) 1	60) 11	60) 9	60) 8	60) 3	60) 2	60) 20	60) 3

KEY

ANSWER KEY

#	Day 91	Day 92	Day 93	Day 94	Day 95	Day 96	Day 97	Day 98	Day 99	Day 100
1)	20	13	12	3	5	9	14	11	17	16
2)	8	12	18	9	12	5	20	18	13	7
3)	7	11	4	11	14	12	10	1	19	9
4)	12	17	19	4	14	0	6	2	14	16
5)	16	11	10	5	19	19	18	16	8	5
6)	3	7	18	4	20	8	19	2	11	7
7)	1	8	20	12	11	1	12	17	3	19
8)	2	14	18	12	20	0	14	0	3	7
9)	5	0	2	19	1	5	10	0	8	16
10)	2	0	10	3	8	14	3	17	5	15
11)	13	2	1	13	10	20	8	9	19	2
12)	8	1	18	15	2	2	12	12	19	7
13)	9	16	4	19	20	16	2	5	14	15
14)	17	9	6	8	7	3	4	0	4	11
15)	16	5	19	8	12	16	20	19	18	1
16)	17	9	14	12	16	1	16	12	0	15
17)	14	14	2	5	13	13	20	7	9	5
18)	18	19	18	8	2	3	5	8	13	10
19)	0	3	15	18	14	10	9	7	15	8
20)	13	12	1	10	8	8	15	19	18	13
21)	13	15	8	6	17	17	9	15	5	15
22)	17	0	9	7	16	7	2	10	6	20
23)	0	6	19	15	2	15	18	10	12	7
24)	11	3	20	17	11	3	20	7	11	17
25)	19	7	10	4	0	17	0	5	20	3
26)	15	14	12	1	19	1	17	9	13	6
27)	10	0	16	2	17	13	8	19	3	18
28)	15	14	17	1	6	7	18	19	4	3
29)	2	4	9	10	4	13	10	11	13	13
30)	7	1	19	10	17	16	5	11	14	6
31)	12	12	4	3	7	10	4	7	8	1
32)	5	2	13	1	4	13	15	13	2	16
33)	14	5	16	19	6	17	1	0	14	19
34)	11	12	10	18	15	1	16	9	15	6
35)	1	6	20	20	4	12	5	16	19	2
36)	12	5	0	15	1	13	1	6	19	1
37)	9	14	2	0	12	6	6	20	5	18
38)	13	5	11	11	17	10	1	14	20	12
39)	19	2	12	1	18	9	4	7	15	2
40)	17	10	3	14	7	14	0	9	0	12
41)	18	18	0	20	20	14	8	5	0	9
42)	12	3	13	8	0	9	8	9	0	8
43)	3	10	17	8	4	10	16	8	7	10
44)	2	13	6	17	15	10	17	4	15	14
45)	16	6	10	7	5	1	0	16	2	15
46)	13	18	4	16	13	14	9	12	17	4
47)	8	10	7	8	20	11	8	4	7	6
48)	7	16	3	5	8	13	17	19	17	16
49)	4	6	16	2	7	6	2	3	14	11
50)	13	16	6	7	10	13	11	3	15	2
51)	9	17	14	10	18	11	16	0	3	3
52)	18	8	3	13	11	9	20	4	8	2
53)	9	20	17	13	5	9	3	1	10	4
54)	11	14	15	12	9	16	10	13	6	0
55)	12	0	4	15	19	3	12	2	20	20
56)	15	18	18	0	14	6	1	3	19	11
57)	15	9	8	9	4	2	10	3	20	4
58)	19	9	9	6	11	1	5	13	8	5
59)	1	9	11	2	5	5	16	10	14	19
60)	3	11	10	1	0	19	10	11	20	5

www.ingramcontent.com/pod-product-compliance
Lightning Source LLC
LaVergne TN
LVHW081411110826
845149LV00010B/1708

* 9 7 8 1 9 6 4 6 9 5 0 2 0 *